Applied Electricity and Electronics
Laboratory Manual

Clair A. Bayne

Electronics Technology Instructor

Venango County Area Vocational-Technical School

Oil City, Pennsylvania

Publisher

The Goodheart-Willcox Company, Inc.

Tinley Park, Illinois

ISBN-13: 978-1-56637-708-9
ISBN-10: 1-56637-708-0

3 4 5 6 7 8 9 10   00   12 11 10 09

Safety in the Electricity/Electronics Laboratory

There is always an element of danger when working with electricity. Observe all safety rules concerning each activity and be particularly careful not to contact any live wire or terminal, even if it is connected to a low voltage. The lab activities do not specify dangerous voltage levels. Keep in mind it is possible at all times to experience a surprising electric shock under certain circumstances. Anyone can be seriously injured by a shock or the results of it. Tools used in the electronics laboratory can cause injury if used carelessly. The importance of following safe practices and procedures cannot be overemphasized. No electronics laboratory is completely safe. Develop sound safety habits!

INTRODUCTION

The wide range of activities in this laboratory manual will aid you in the study of electricity and electronics. While working with circuits, you will use test instruments and electronic components to explore how circuits operate. Keep in mind that measurement results are affected by the type and calibration of test equipment, as well as the characteristics of the circuits being tested. Instructors should allow for these minor differences when checking student measurements.

In this "information age," you need to know how to communicate your ideas to others. In many of the activities, you are directed to write your observations. This calls on you to form ideas or draw comparisons of the data gathered and to communicate that information to others.

The overall objectives of the lab activities are to practice:

- Reading schematic diagrams and construction of circuits from diagrams
- Measuring with meters and other test equipment
- Developing a better understanding of how circuits operate
- Proving electrical rules or laws
- Investigating electronic circuits

Before performing the activities, study the corresponding textbook chapter. When used in conjunction with the textbook and other reference material, this lab manual will make your study of electricity and electronics both interesting and useful.

Each activity is divided into the following sections:

Discussion. Gives additional information about the subject or reviews material from the textbook.

Objectives. States the learning objectives of the lab experiment and the skills to be mastered.

Materials and Equipment. Lists the components and equipment needed to perform the activity.

Procedures. Outlines the specific steps to follow for the lab to work properly. If the steps are completed out of order or skipped, the outcome of the experiment may not be correct. Also, carrying out proper procedures will help you develop the good habit of following directions. Sometimes questions and answers replace the procedures section.

Summing Up. Keep a notebook for recording any observations, problems, and conclusions for each activity. An observation is a record of what happened. For example, maybe resistor X was hot compared to resistor Y. Perhaps voltage Z was much lower or higher than expected, or the output voltage was zero.

Record any problems, such as difficulty connecting the circuit, wrong settings, equipment problems, or loss of power. Show why the lab did not work.

Conclusions are your written opinion of the experiment based on the observations, problems, and results. This is where you explain *why* things happened. Questions in the experiments may ask you to compare data to assist you in forming a conclusion.

I hope you enjoy the activities and, at the same time, increase your interest and understanding of electricity and electronics. It is an exciting field that is full of opportunity for you.

Clair A. Bayne

CONTENTS

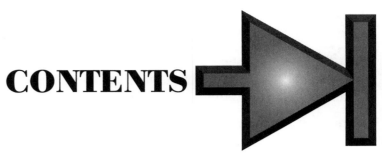

Section I: Fundamentals of Direct Current

Section II: Electronic Assembly

Section III: Fundamentals of Alternating Current

Section IV: Inductance, Capacitance, and RCL Circuits

Section V: Ac Power and Motors

Section VI: Electronic Principles

FUNDAMENTALS OF ELECTRICITY

Activity 1-1:
Lab Safety Inspection

Name _____

Date _____

Class _____ Score _____

Discussion

Safety is a major consideration for an employer. If employees are injured on the job, it costs the employer money in terms of absent workers and increased workman's compensation insurance fees. The key to maximum employee output is a safe workplace.

Objectives

In this activity, you will:
• Inspect the lab area for safety violations.
• Become familiar with the location of emergency equipment.

Materials and Equipment

1–Pen or pencil

Procedures

Inspect the laboratory for safety violations. Put a check mark next to the items that pass inspection.

_____ 1. Cover plates on all ac outlets.

_____ 2. Guards on power equipment.

_____ 3. Power cords on electrical equipment in good condition. Check ground conductor.

_____ 4. Fire extinguishers inspected on _____ (date).

_____ 5. Emergency power disconnects operational.

_____ 6. Fire exit diagram displayed.

_____ 7. Circuit breaker box accessible.

_____ 8. Emergency telephone numbers readily available.

_____ 9. Extension cords in proper working condition.

_____10. Other (explain).

_____

_____

(Continued)

Summing Up

In your notebook, record any observations, problems, and conclusions for this activity.

Activity 1-2:
Construction of Cells

Name _____

Date _____

Class _____ Score _____

Discussion

Electricity comes from several different sources, including chemical, magnetic, heat, and solar (light) energy. A common chemical source is the cell, often referred to as a battery. A battery is made up of two or more cells as shown in **Figure 1-2A**. Chemical cells operate because of the chemical reaction of a solution known as an electrolyte and two dissimilar (different) metals. This type of cell is called a wet cell. A cell in which the electrolyte is absorbed by paper or formed into a paste is called a dry cell.

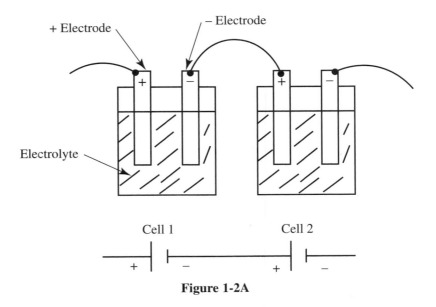

Figure 1-2A

Objectives

In this activity, you will:
- Construct a wet cell.
- Build a two cell battery using lemons as the electrolyte.

Materials and Equipment

1–VOM
1–Plastic container (35 mm film or similar)
3–Galvanized nails, size #6 (zinc-coated)
3–Printed circuit board pieces, 1/2″ × 2″ (for the copper foil)
2–Lemons
1/2 tsp. table salt

Procedures

For the *wet cell,* see **Figure 1-2B.**

1. Carefully cut the top of the plastic container so the nail and PC board fit snugly.

(Continued)

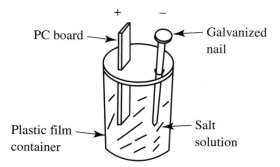

Figure 1-2B

2. Fill the container about 3/4 full with warm water. Stir the salt in until it is dissolved.

3. Attach the lid to the container, making sure the nail and PC board are down into the liquid.

4. Using a voltmeter, measure the voltage between the nail and PC board.

   Volts = _____

For the *lemon cell,* see **Figure 1-2C.**

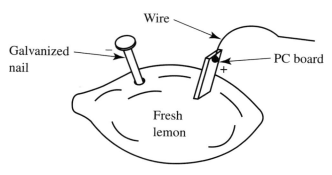

Figure 1-2C

1. Clean the printed circuit board.

2. Prepare the PC board pieces by soldering a wire on one end of the board. If soldering is not possible, drill a hole and use a machine screw and nut to attach the wire.

3. Cut a slot into each of the lemons for the PC board.

4. Push the PC board and nail into each lemon.

5. Measure the voltage between the PC board and nail.

   Lemon 1 = _____ V

   Lemon 2 = _____ V

6. As shown in **Figure 1-2D,** connect the two cells together and measure the voltage again.

   Volts = _____

7. How does the voltage of one cell compare with the voltage of the two cells connected together?

   _____

   _____

(Continued)

Name _____

8. How does the lemon cell voltage compare to the salt water cell?

_____

_____

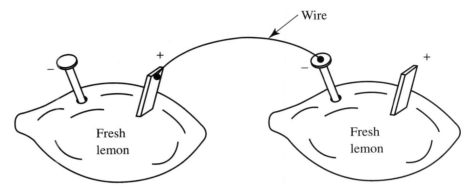

Figure 1-2D

Summing Up

In your notebook, record any observations, problems, and conclusions for this activity.

<table>
<tr><td>**Activity 1-3:**
An Electrical Circuit</td><td>Name _____
Date _____
Class _____ Score _____</td></tr>
</table>

Discussion

Different electrical circuits are used when working with electricity. Although the circuit you will build is simple, it demonstrates various basic principles of electricity.

Objective

In this activity, you will build a simple electrical circuit.

Materials and Equipment

1–6 V battery, Ray-O-Vac #945 or similar
1–Breadboard
2–Light-emitting diodes, XC556RB or similar
Miscellaneous wire
1–Resistor, 150 Ω
1–Switch (knife switch preferred)
1–Voltmeter

Procedures

1. Connect the circuit as shown in **Figure 1-3.**

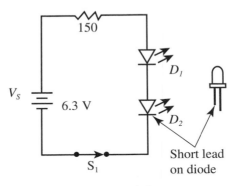

Figure 1-3

2. Open and close the switch, and observe the results.

3. Have your instructor assist in measuring the voltage across each bulb.

   Voltage across LED 1 = _____ V

   Voltage across LED 2 = _____ V

   Voltage across the resistor = _____ V

4. Measure the voltage of the source.

   Source voltage = _____ V

(Continued)

5. How does the sum of the voltages across the LEDs compare to the source voltage?

_____

_____

Summing Up

In your notebook, record any observations, problems, and conclusions for this activity.

MATH FOR ELECTRICITY

Activity 2-1:
Fractions and Reciprocals

Name _____

Date _____

Class _____ Score _____

Discussion

Many problems in electricity involve voltages or currents that contain fractions. Solving these problems will depend on your ability to use fractions. If the numerator and denominator of a fraction are multiplied by the same amount, the value of the fraction is unchanged, for example:

$$\frac{3}{4} \times \frac{2}{2} = \frac{6}{8}$$

Objective

In this activity, you will practice changing fractions to decimals.

Materials and Equipment

1–Pen or pencil

Procedures

Change the following fractions to decimal numbers.

1. a. $\dfrac{3}{6}$

   b. $\dfrac{4}{16}$

   c. $\dfrac{5}{30}$

   d. $\dfrac{7}{35}$

   e. $\dfrac{6}{48}$

   f. $\dfrac{6}{9}$

2. a. $\dfrac{2}{10}$

   b. $\dfrac{5}{15}$

   c. $\dfrac{7}{21}$

1. a. _____

   b. _____

   c. _____

   d. _____

   e. _____

   f. _____

2. a. _____

   b. _____

   c. _____

(Continued)

d. $\dfrac{9}{12}$ d. _____

e. $\dfrac{8}{5}$ e. _____

f. $\dfrac{8}{20}$ f. _____

Find the reciprocals of the following numbers.

3. a. 6   3. a. _____

    b. 12   b. _____

    c. 5   c. _____

    d. 8   d. _____

    e. 10   e. _____

    f. 60   f. _____

4. a. 1000   4. a. _____

    b. 500   b. _____

    c. 800   c. _____

    d. 6000   d. _____

    e. 1200   e. _____

    f. 0.6   f. _____

Activity 2-2:	Name _____
Powers and Roots	Date _____
	Class _____ Score _____

Discussion

The ability to solve powers and roots, and to operate a calculator are essential skills for working with electrical circuits.

Objective

In this activity, you will practice solving powers and roots of numbers.

Materials and Equipment

1–Pen or pencil
1–Calculator

Procedures

Solve the following powers.

1. a. 5^2 1. a. _____
   b. 8^4 b. _____
   c. 3^3 c. _____
   d. 4^5 d. _____
   e. 3^8 e. _____
   f. 8^3 f. _____

2. a. 18^2 2. a. _____
   b. 66^2 b. _____
   c. 29^3 c. _____
   d. 30^4 d. _____
   e. 25^2 e. _____
   f. 32^3 f. _____

3. a. 45^2 3. a. _____
   b. 4^2 b. _____
   c. 60^2 c. _____
   d. 10^2 d. _____
   e. 100^2 e. _____
   f. 15^2 f. _____

Solve the following roots.

4. a. $\sqrt{144}$ 4. a. _____
   b. $\sqrt{16}$ b. _____
   c. $\sqrt{81}$ c. _____

(Continued)

      d. $\sqrt{625}$

      e. $\sqrt{676}$

      f. $\sqrt{900}$

5. a. $\sqrt[6]{46656}$

      b. $\sqrt[4]{625}$

      c. $\sqrt[3]{729}$

      d. $\sqrt[9]{512}$

      e. $\sqrt[5]{243}$

      f. $\sqrt[4]{20736}$

6. a. $\sqrt{100}$

      b. $\sqrt[3]{125000}$

      c. $\sqrt{50}$

      d. $\sqrt{69.6}$

      e. $\sqrt{27.7729}$

      f. $\sqrt[5]{7776}$

d. _____

e. _____

f. _____

5. a. _____

   b. _____

   c. _____

   d. _____

   e. _____

   f. _____

6. a. _____

   b. _____

   c. _____

   d. _____

   e. _____

   f. _____

Activity 2-3:
Adding Signed Numbers

Name _____

Date _____

Class _____ Score _____

Discussion ━━━━━━━━━━━━━━━━━━━━━━━━━━━━━━━━━━━━━━━

Solving circuit problems involve voltages or currents with a positive or negative polarity. Therefore, the addition of signed numbers is used extensively when working with electrical circuits.

Objective ━━━━━━━━━━━━━━━━━━━━━━━━━━━━━━━━━━━━━━━

In this activity, you will practice adding signed numbers.

Materials and Equipment ━━━━━━━━━━━━━━━━━━━━━━━━━━

1–Pen or pencil

Procedures ━━━━━━━━━━━━━━━━━━━━━━━━━━━━━━━━━━━━━

Add the following signed numbers.

1. a.　5
　　　　3

b.　−8
　　+4

c.　−3
　　+5

　 d.　−4
　　　−7

e.　−3
　　　6

f.　8
　　6

2. a.　−8
　　　+1

b.　+6
　　−5

c.　−9
　　−7

　 d.　−4
　　　　6

e.　−3
　　+3

f.　−2
　　+8

3. a.　−3
　　　−6

b. −10
　　−11

c.　21
　　−5

　 d. −10
　　　14

e.　−2
　　−13

f. −16
　　+16

4. a.　−1
　　　−12

b.　14
　　−14

c.　　2
　　−13

　 d.　40
　　　−1

e. −40
　　−1

f.　−8
　　−8

(Continued)

5. a.   5   b.   6   c. −9
        −3   −7   +9
        −4   −8   7

    d.   4   e. −10   f.   7
        −6   6   −2
         7   +10   +5

Activity 2-4:	Name _____
Equation Operations	Date _____
	Class _____ Score _____

Discussion

Many mathematical problems in electricity involve equations with unknowns. Getting the unknown where it should be is a basic skill in performing equation operations.

Objective

In this activity, you will change or rearrange an equation for an unknown.

Materials and Equipment

1–Pen or pencil

Procedures

Find the unknowns in the following equations.

1. $L = RTZ$ $T =$_____

2. $S = \dfrac{Q}{VT}$ $T =$_____

3. $W = \dfrac{BT}{L^2}$ $L =$_____

4. $T = INR$ $R =$_____

5. $k = \dfrac{EPT^3}{G}$ $P =$_____

6. $P = \dfrac{U^2}{TYQ}$ $Y =$_____

**Activity 2-5:
Solving Proportions**

Name _____

Date _____

Class _____ Score _____

Discussion

Electrical problems involving transformers and other devices that depend on a set of ratios can be solved using proportions. Solving proportions requires basic equation skills. A proportion is a statement of equality between two ratios. Remember the rule: In cross-multiplication, the two products must be equal. If they are not equal, the proportion is invalid.

Objective

In this activity, you will solve for the unknown in proportion equations.

Materials and Equipment

1–Pen or pencil

Procedures

Find the unknown in the following equations.

1. $P = 800$, $J = 200$, and $Q = 30$. Solve for L.   1. _____

$$\frac{P}{L} = \frac{J}{Q}$$

2. $X = 1$, $N = 50$, and $T = 0.5$. Solve for B.   2. _____

$$\frac{X}{N} = \frac{T}{B}$$

3. $B = 10$, $C = 12$, and $D = 40$. Solve for A.   3. _____

$$\frac{A}{B} = \frac{C}{D}$$

4. $V = 6$, $W = 20$, and $Z = 100$. Solve for Y.   4. _____

$$\frac{V}{W} = \frac{Y}{Z}$$

Activity 2-6: Scientific and Engineering Notation

Name _____

Date _____

Class _____ Score _____

Discussion ━━━━━━━━━━━━━━━━━━━━━━━━━━━━━━━━━━

Like other scientific fields, electricity and electronics use scientific and engineering notation to express numbers. Such notation makes working with very large or small numbers easier, particularly on a calculator.

Objective ━━━━━━━━━━━━━━━━━━━━━━━━━━━━━━━━━━

In this activity, you will express numbers in scientific and engineering notation.

Materials and Equipment ━━━━━━━━━━━━━━━━━━━━━━━━

1–Pen or pencil

Procedures ━━━━━━━━━━━━━━━━━━━━━━━━━━━━━━━━━━

1. Express the following numbers in scientific notation.

   a. 1,900,000

   b. 357,000

   c. 380

   d. 2,500

   e. 3,700,000

   f. 0.00527

   g. 0.0012

   h. 0.0000075

   i. 0.00019

   j. 0.0000017

   1. a. _____
   b. _____
   c. _____
   d. _____
   e. _____
   f. _____
   g. _____
   h. _____
   i. _____
   j. _____

2. Express the following numbers in engineering notation.

   a. 2,400,000

   b. 222,000

   c. 960

   d. 1,800

   e. 6,500,000

   f. 0.00725

   g. 0.0018

   h. 0.0000033

   i. 0.00092

   j. 0.0000071

   2. a. _____
   b. _____
   c. _____
   d. _____
   e. _____
   f. _____
   g. _____
   h. _____
   i. _____
   j. _____

CONDUCTORS, INSULATORS, AND RESISTORS

3

Activity 3-1:
Preparation of Wires

Name _____

Date _____

Class _____ Score _____

Discussion

Many pieces of wire are needed to construct a circuit. The wires should not have insulation damage, and the conductor should not be nicked or scraped. A nicked wire can break when it is bent to make a connection, leaving a small piece of wire in the breadboard. This would make future connections impossible. If the stripped portion of the wire is too long, it may come in contact with another conductor and cause a short circuit. In the real world, such defects can cause major equipment problems. An entire assembly line could be stopped, costing thousands of dollars in down time. A defect in a guided missile could cause loss of life, aircraft, or ships.

Objective

In this activity, you will cut 10 pieces of wire to proper length and strip each end 1/4″ (6 mm).

Materials and Equipment

22-gage wire
Diagonal wire-cutting pliers
Inch ruler
Metric ruler
Scissors
Wire stripper

Procedures

1. Make sure you understand and follow safety precautions for using wire strippers and wire cutters.

2. Using an inch ruler or a metric ruler, measure and cut the wire to each of the following sizes.

   a. 4″

   b. 2 3/4″

   c. 12.5 cm

   d. 1 1/8″

   e. 8.3 cm

(Continued)

      f. 5.5 cm

      g. 15 cm

      h. 4.7 cm

      i. 3 5/16″

      j. 6 1/2″

3. Strip 1/4″ (6 mm) of insulation from each end.

4. Use masking or clear tape to attach the wires to a sheet of strong paper or lightweight cardboard. Turn it in to your instructor for evaluation.

Summing Up

In your notebook, record any observations, problems, and conclusions for this activity.

Activity 3-2:
Resistor Identification

Name _____

Date _____

Class _____ Score _____

Discussion ──────────────────────────

Many resistors often are required in the construction of circuits. Using a resistor with an incorrect value will result in a circuit that does not operate correctly. You should be able to identify resistors quickly so valuable lab time is not lost.

Objective ──────────────────────────

In this activity, you will identify resistors using the resistor color code.

Materials and Equipment ──────────────────

1–Pen or pencil

Procedures ──────────────────────────

1. Give the numerical value for each color.

   Blue _____   Yellow _____

   Black _____   Gray _____

   Red _____   White _____

   Green _____   Violet _____

   Orange _____   Brown _____

2. Give the value and tolerance of each of the resistors.

   a. Gray, Red, Orange, Gold   Value _____   Tol. _____

   b. Yellow, Violet, Red, Silver   Value _____   Tol. _____

   c. Brown, Green, Black, Gold   Value _____   Tol. _____

   d. White, Brown, Brown, Silver   Value _____   Tol. _____

3. Convert the following color coded resistors to their proper value.

   a. Red, Red, Blue, Silver _____

   b. Orange, White, Red _____

   c. Gray, Red, Black, Gold _____

   d. Brown, Black, Red, Silver _____

   e. Orange, Orange, Green _____

   f. Yellow, Violet, Yellow, Gold _____

   g. Red, Violet, Gold, Silver _____

   h. Brown, Gray, Silver _____

   i. Green, Blue, Red, Silver _____

(Continued)

4. Convert the following values to the proper color codes.

	1st Band	2nd Band	3rd Band	Tolerance
a. 18,000 10%				
b. 47,000 5%				
c. 0.82 5%				
d. 3300 5%				
e. 5,600,000 10%				
f. 120,000 10%				
g. 4700 5%				
h. 12 5%				
i. 10,000 10%				
j. 8200 5%				
k. 3.3 5%				

Summing Up

In your notebook, record any observations, problems, and conclusions for this activity.

ELECTRICAL CIRCUITS

Activity 4-1:
Node Rule

Name _____

Date _____

Class _____ Score _____

Discussion

When working with or analyzing a circuit, consider the circuit current(s). This will increase your understanding of circuit action and speed up circuit analysis. Remember the node rule: The current leaving a node must be equal to the current entering a node. The solution of node current(s) is a matter of adding or subtracting the current(s) involved.

Objective

In this activity, you will use the current node rule to solve for the current flowing in a circuit.

Materials and Equipment

1–Pen or pencil

Procedures

Using addition or subtraction, find the unknown currents in **Figures 4-1A** through **4-1G.**

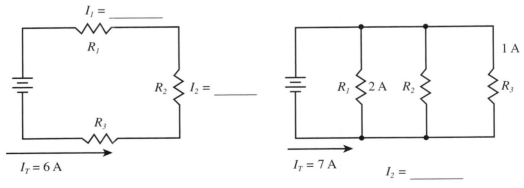

Figure 4-1A   Figure 4-1B

(Continued)

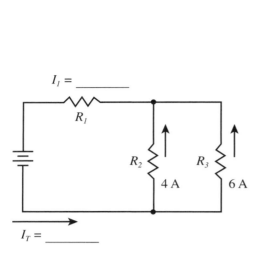

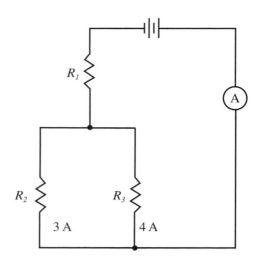

Figure 4-1C

Figure 4-1D

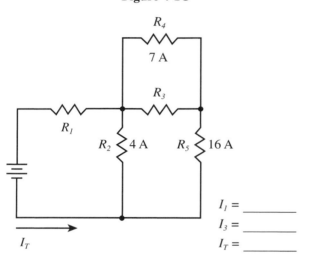

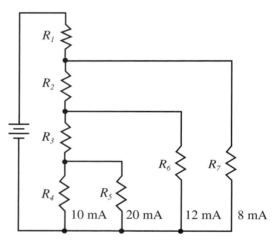

Figure 4-1E

Figure 4-1F

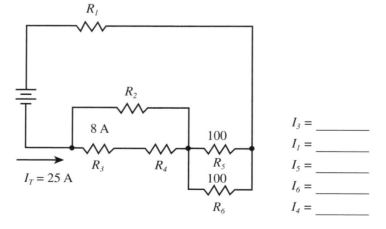

Figure 4-1G

Summing Up

In your notebook, record any observations, problems, and conclusions for this activity.

Activity 4-2:	Name _____
Ohm's Law	Date _____
	Class _____ Score _____

Discussion

Calculating current, resistance, or voltage is often necessary when working with electricity. Ohm's law is used to accomplish the calculations. Place the known values into the equation and allow the answer to the unknown to "fall out."

Example: Find the current if the voltage is 12 V and the resistance is 6 Ω.

$V = IR$

$12 = I\,6$ (Divide both sides by 6.)

$2 = I$

Objective

In this activity, you will solve for the unknowns using Ohm's law.

Materials and Equipment

1–Pen or pencil

Procedures

Using Ohm's law, find the unknown in **Figures 4-2A** through **4-2F.**

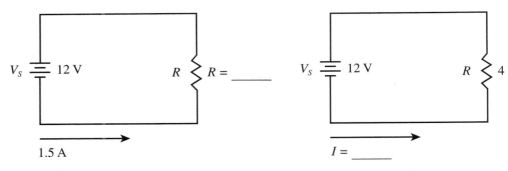

Figure 4-2A   **Figure 4-2B**

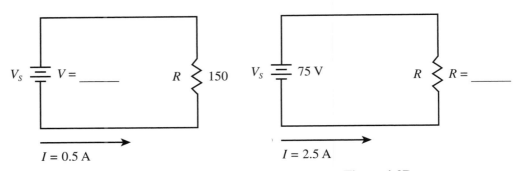

Figure 4-2C   **Figure 4-2D**

(Continued)

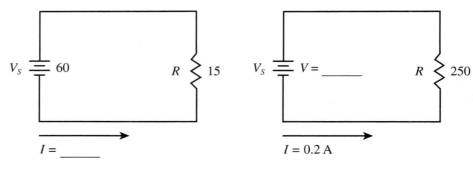

Figure 4-2E   **Figure 4-2F**

Summing Up

In your notebook, record any observations, problems, and conclusions for this activity.

Activity 4-3:	Name _____
Basic Circuit	Date _____
Construction	Class _____ Score _____

Discussion ────────────────────────────

During your electronic training, you will build resistor, transistor, digital, and other types of circuits. Rapid reading of circuit diagrams and proper circuit construction are fundamental skills to master. Furthermore, you must thoroughly understand the equipment used to build and test these circuits.

Objective ────────────────────────────

In this activity, you will build basic electrical circuits.

Materials and Equipment ────────────────────────────

1–Breadboard
1–Power supply
5–Resistors (any value)

Procedures ────────────────────────────

Using resistors of any value, build the circuits shown in **Figures 4-3A** through **4-3D.** Your instructor will check each circuit for proper connections and polarity of the voltage source.

Caution! Do not connect the circuits to any active power source. This is an exercise in *connecting* circuits. No power is necessary.

Circuit	1	2	3	4
Instructor check	____	____	____	____

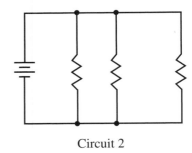

Circuit 1   Circuit 2

Figure 4-3A   **Figure 4-3B**

(Continued)

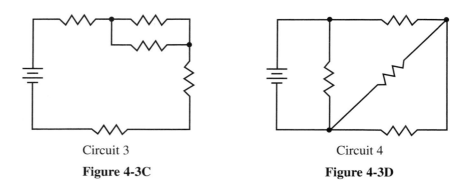

Circuit 3   Circuit 4

Figure 4-3C   **Figure 4-3D**

Summing Up

In your notebook, record any observations, problems, and conclusions for this activity.

Activity 4-4:
Conductance

Name _____

Date _____

Class _____ Score _____

Discussion

Calculating the resistance of parallel circuits is often necessary when working with electricity. There are several methods for finding the total resistance. For parallel circuits, it is calculated using conductance. In Chapter 5 you will use a meter to measure resistance.

Objective

In this activity, you will calculate the conductance and resistance of circuits.

Materials and Equipment

1–Pen or pencil

Procedures

Using conductance, find the unknowns in the **Figures 4-4A** through **4-4D.**

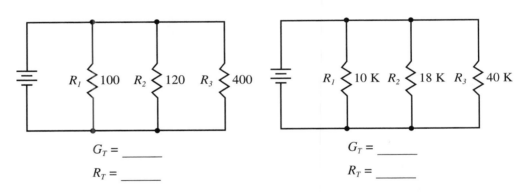

$G_T =$ _____

$R_T =$ _____

Figure 4-4A

$G_T =$ _____

$R_T =$ _____

Figure 4-4B

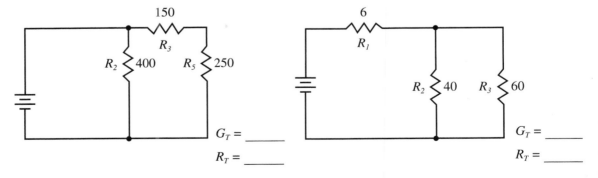

$G_T =$ _____

$R_T =$ _____

Figure 4-4C

$G_T =$ _____

$R_T =$ _____

Figure 4-4D

Summing Up

In your notebook, record any observations, problems, and conclusions for this activity.

USING ELECTRICAL METERS

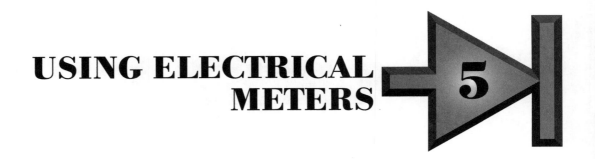

Activity 5-1:	Name _____
Measuring Resistance	Date _____
	Class _____ Score _____

Activity 5-1:
Measuring Resistance

Discussion

Using a meter to measure resistance is basic to the field of electricity and electronics. This lab will also reinforce the ability to read the resistor color codes.

Objectives

In this activity, you will:
- Measure the resistance of resistors using a VOM.
- Compare color code values with measured values.

Materials and Equipment

One each of the following resistors:

100 Ω	22,000 Ω
470 Ω	33,000 Ω
1,000 Ω	68,000 Ω
3,900 Ω	470,000 Ω
4,700 Ω	1,000,000 Ω

1–Volt-ohm-milliammeter (VOM)

Procedures

1. Obtain the resistors listed in the table in **Figure 5-1.**

2. Using the resistor color codes, write the values of the resistors on the left side of the table.

3. Using a meter, measure the resistance of each resistor. Record the measurements on the right side of the table.

4. Describe the differences between the color code values and the measured values.

   _____

   _____

5. Indicate if any of the resistors are out of tolerance.

   _____

   _____

(Continued)

Color Code Value	Color Code	Measured Value
	Yellow–Violet–Brown	
	Red–Red–Orange	
	Brown–Black–Brown	
	Yellow–Violet–Red	
	Blue–Gray–Orange	
	Brown–Black–Green	
	Yellow–Violet–Yellow	
	Brown–Black–Red	
	Orange–Orange–Orange	
	Orange–White–Red	

Figure 5-1

Summing Up

In your notebook, record any observations, problems, and conclusions for this activity.

**Activity 5-2:
Measuring Circuit
Resistance**

Name _____

Date _____

Class _____ Score _____

Discussion

Resistance measurements can verify whether a circuit contains a short or open circuit. **Caution! Before making any resistance measurement on a circuit, be sure the power to the circuit is turned off.**

Objective

In this activity, you will build the three basic electric circuits and measure the total resistance of each.

Materials and Equipment

One each of the following resistors:
150 Ω
1000 Ω
3300 Ω
1–VOM

Procedures

1. Connect the circuit shown in **Figure 5-2A** and measure the total resistance.

   $R_T =$ _____

2. How does the total resistance measurement compare to the calculated total resistance?

   _____

   _____

   _____

3. How does the measurement relate to the rule for the total resistance of a series circuit?

   _____

   _____

   _____

4. Connect the circuit shown in **Figure 5-2B** and measure the total resistance.

   $R_T =$ _____

5. How does the total resistance measurement compare to the calculated total resistance?

   _____

   _____

   _____

(Continued)

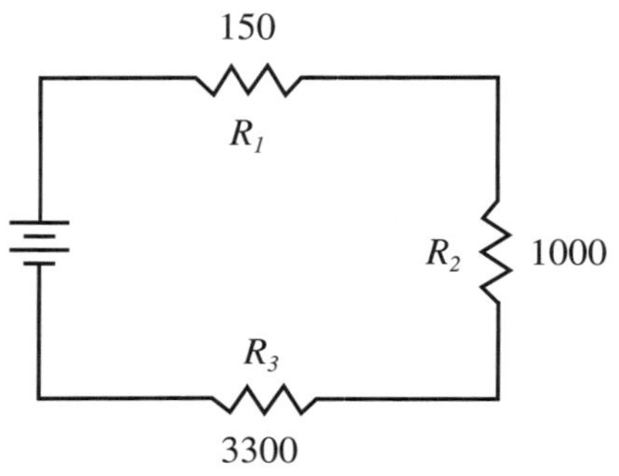

Figure 5-2A

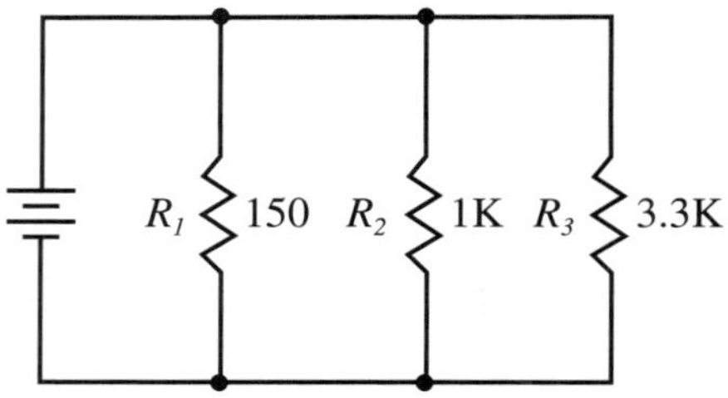

Figure 5-2B

6. How does the measurement relate to the rule for the total resistance of a parallel circuit?

_____

_____

_____

7. Connect the circuit of **Figure 5-2C** and measure the total resistance.

$R_T =$ _____

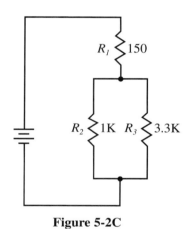

Figure 5-2C

8. How does the total resistance measurement compare to the calculated total resistance?

_____

_____

_____

9. How does the measurement relate to the rule for the total resistance of a series-parallel circuit?

_____

_____

_____

Summing Up ─────────────────────────────────

In your notebook, record any observations, problems, and conclusions for this activity.

Activity 5-3:
Testing for Continuity

Name _____

Date _____

Class _____ Score _____

Discussion

The continuity test is used every day by technicians. The test verifies a circuit has a complete path for current to flow. It is the simplest test for providing information about the condition of a component or circuit.

Objective

In this activity, you will test components and circuits for electrical continuity.

Materials and Equipment

Assorted fuses
Assorted switches
Wire cable (more than two conductors)
1–Coil
1–Transformer
Analog or digital meter

Procedures

Several components are listed in the table in **Figure 5-3.** Test each one for continuity, and record the results for good, open, and shorted conditions.

Component	Good	Open	Shorted
Switch 1			
Fuse 1			
Switch 2			
Cable			
Switch 3			
Fuse 2			
Coil			
Switch 4			
Transformer			

Figure 5-3

Summing Up

In your notebook, record any observations, problems, and conclusions for this activity.

Activity 5-4:
Measuring Voltages

Name _____

Date _____

Class _____ Score _____

Discussion

A voltage measurement indicates either a voltage source or voltage drop. Voltage sources can be connected in series or in parallel. In series, the total voltage is the sum of the various voltages in the circuit. The total voltage depends on the direction (polarity) the sources are connected. In parallel, the final voltage is the same as one of the sources.

Objectives

In this activity, you will:
- Measure voltages using an analog or digital voltmeter.
- Investigate series and parallel battery connections.
- Demonstrate the effects of connecting voltage sources in series and parallel with different polarities.

Materials and Equipment

Assorted wire
3–6 V batteries
1–9 V transistor battery
1–Dc power supply
1–VOM

Procedures

1. In the table in **Figure 5-4A,** record the voltage indicated on each battery.

2. Measure and record the voltage of each battery.

Battery	Listed Voltage	Measured Voltage	Condition
B_1			
B_2			
B_3			
B_4 (9 V)			

Figure 5-4A

3. What do the readings indicate about the condition of each battery?

_____

_____

_____

_____

(Continued)

4. Connect the circuits shown in **Figure 5-4B** and record the voltages.

$V_1 = \underline{\hspace{1.5in}}$

$V_2 = \underline{\hspace{1.5in}}$

$V_3 = \underline{\hspace{1.5in}}$

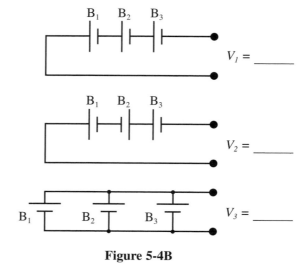

$V_1 = \underline{\hspace{0.8in}}$

$V_2 = \underline{\hspace{0.8in}}$

$V_3 = \underline{\hspace{0.8in}}$

Figure 5-4B

5. What are the differences between V_1 and V_2? Why?

_____

_____

_____

6. What are the differences between V_1 and V_3? Why?

_____

_____

_____

7. Adjust the variable dc power supply for each voltage listed in the table in **Figure 5-4C.** Have your instructor verify each reading.

Voltage	Instructor Check
5	
2	
0.5	
8.4	
3.6	
14	

Figure 5-4C

Summing Up

In your notebook, record any observations, problems, and conclusions for this activity.

Activity 5-5:	Name _____
Measuring Current	Date _____
	Class _____ Score _____

Discussion

Measuring current is a skill technicians apply to various types of circuit conditions. When you are measuring current, the meter *must be in series* with the circuit being measured. Placing a current meter in parallel could damage the meter beyond repair.

Objective

In this activity, you will measure current with an analog or digital voltmeter.

Materials and Equipment

One each of the following resistors:
   100 Ω
   560 Ω
   220 Ω
1–Dc power supply
1–VOM

Procedures

1. Construct the series circuit shown in **Figure 5-5A.**

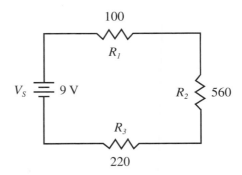

Figure 5-5A

2. Make and record the required current measurements in the table in **Figure 5-5B.** **Caution! A current meter must be connected *in series* with the current.**

I_{R_1}	I_{R_2}	I_{R_3}	I_T

Figure 5-5B

(Continued)

3. How do the measurements compare to the series circuit rules in Chapter 4?

_____

_____

4. Construct the parallel circuit shown in **Figure 5-5C.**

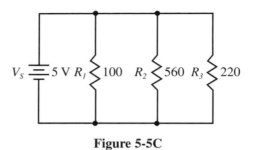

Figure 5-5C

5. Make and record the required current measurements in the table in **Figure 5-5D.**

I_{R_1}	I_{R_2}	I_{R_3}	I_T

Figure 5-5D

6. How do the measurements compare to the parallel circuit rules in Chapter 4?

_____

_____

Summing Up ——————————————————————————

In your notebook, record any observations, problems, and conclusions for this activity.

ELECTRICAL QUANTITIES

6

Activity 6-1:
Wire Resistance and
Converting Units of
Measurement

Name _____

Date _____

Class _____ Score _____

Discussion

In many cases, an electrical measurement will be given, but it may be needed in another form. This exercise will provide practice converting units of measurement.

Objective

In this activity, you will solve wire resistance problems and convert units of measurement.

Materials and Equipment

1–Calculator (optional)
1–Light-dependent resistor
1–Pen or pencil

Procedures

1. Measure the resistance of a light-dependent resistor. Does it have a positive or negative light coefficient? Explain.

   _____

   _____

   _____

2. What is the resistance of 125′ of #26 copper wire at 30° C?

   _____

3. What is the resistance of 125′ of #26 aluminum wire at 30° C?

   _____

4. Convert the following units.

   a. 470 mH = _____ H

   b. 0.028 A = _____ mA

   c. 10,000 pF = _____ µF

(Continued)

51

    d. 0.003 S=_____ mS

    e. 800 pA =_____ μA

    f. 1500 V =_____ kV

    g. 150 pF =_____ μF

    h. 3300 Ω =_____ kΩ

Summing Up

In your notebook, record any observations, problems, and conclusions for this activity.

ELECTRICAL POWER 7

Activity 7-1:
Resistor Power Rating

Name _____

Date _____

Class _____ Score _____

Discussion

If the power rating of a resistor is too low, the resistor will become overheated. This can cause a change in its resistance, damage nearby components, or start a fire. Always replace a resistor with one of the same or higher power rating.

Objective

In this activity, you will investigate the power rating of resistors.

Materials and Equipment

2–47 Ω resistors
1–68 Ω resistor
1–150 Ω resistor/5 W
1–Breadboard
1–VOM

Procedures

1. Calculate the total resistance and total current.

   $R_T =$ _____

   $I_T =$ _____

2. Calculate the power of each of the resistors in **Figure 7-1A,** and record the calculations in the table in **Figure 7-1B.**

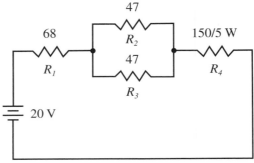

Figure 7-1A

(Continued)

P_{R_1}	P_{R_2}	P_{R_3}	P_{R_4}	P_T

Figure 7-1B

4. Construct the circuit shown in Figure 7-1A.

5. Turn the power to the circuit and wait two minutes. This allows the resistors to come up to their operating temperature.

6. Investigate the temperature of each of the resistors. **Warning! Some resistors can be hot enough to produce a burn.**

7. Write a summary of your findings.

_____

_____

_____

_____

_____

_____

_____

_____

_____

_____

Summing Up

In your notebook, record any observations, problems, and conclusions for this activity.

DC CIRCUIT ANALYSIS 8

Activity 8-1:
Verifying the Series
Voltage Rule

Name _____

Date _____

Class _____ Score _____

Discussion

The voltage rule states: The sum of the voltage drops around a circuit is equal to the source voltage. Kirchhoff's law states: The algebraic sum of all the voltages around a series circuit, including the source, equals zero.

Objective

In this activity, you will verify the series voltage rule and Kirchhoff's law.

Materials and Equipment

One each of the following resistors:
　　100 Ω
　　150Ω
　　220Ω
　　560 Ω
　　820 Ω
1–Breadboard
1–Power supply
1–VOM

Procedures

1. Construct the circuit shown in **Figure 8-1A.**

2. While keeping the polarity of the test probes in the same direction around the circuit, measure the voltage across each resistor and the source.

3. Record the measurements in the table in **Figure 8-1B.**

4. Add the voltages to find the total.

5. How do the measurements compare with the results of the series rule and Kirchhoff's law?

_____

_____

_____

_____

(Continued)

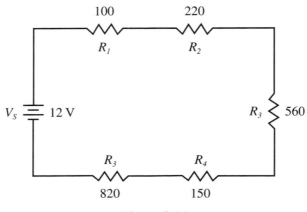

Figure 8-1A

R	V drop
R_1	
R_2	
R_3	
R_4	
R_5	

Figure 8-1B

Summing Up

In your notebook, record any observations, problems, and conclusions for this activity.

Activity 8-2:
Verifying the Parallel
Voltage Rule

Name _____

Date _____

Class _____ Score _____

Discussion

The voltage rule states: The voltage across a parallel circuit is the same or equal. This activity will prove or verify the parallel voltage rule.

Objective

In this activity, you will verify the parallel voltage rule.

Materials and Equipment

One each of the following resistors:
- 560 Ω
- 820 Ω
- 1500 Ω
- 2200 Ω

1–Breadboard
1–Power supply
1–VOM

Procedures

1. Construct the circuit shown in **Figure 8-2A.**

2. Measure the voltage across each resistor and the source. Record the measurements in the table in **Figure 8-2B.**

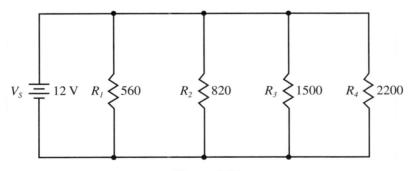

R	V drop
R_1	
R_2	
R_3	
R_4	

Figure 8-2A   Figure 8-2B

3. How do the measurements compare to the parallel voltage rule?

_____

_____

_____

_____

(Continued)

Summing Up

In your notebook, record any observations, problems, and conclusions for this activity.

Activity 8-3:
Open Circuit Effects

Name _____

Date _____

Class _____ Score _____

Discussion

At times a circuit will not operate because of an open somewhere in the current path. Although continuity testing can locate an open circuit, the power must be turned off. By taking voltage measurements at points within the circuit, you can locate the open without this inconvenience.

Objective

In this activity, you will investigate the effects of an open circuit on voltage measurements.

Materials and Equipment

One each of the following resistors:
   150 Ω
   470 Ω
   680 Ω
   1.5K Ω
   2.2K Ω
1–Breadboard
1–Power supply
1–VOM

Procedures

1. Construct the circuit shown in **Figure 8-3A.**

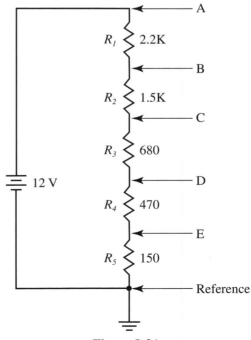

Figure 8-3A

(Continued)

2. Keeping the negative probe at the ground point, measure and record the voltage at each of the points in the table in **Figure 8-3B.**

Point	Closed Circuit Voltage	Open Circuit Voltage
A		
B		
C		
D		
E		

Figure 8-3B

3. Open the circuit at point C in **Figure 8-3C.**

4. Keeping the negative probe at ground point, test for voltage at each of the points and record them in the table.

5. Measure at the different points at C in **Figure 8-3D.**

$V_1 =$ _____

$V_2 =$ _____

$V_3 =$ _____

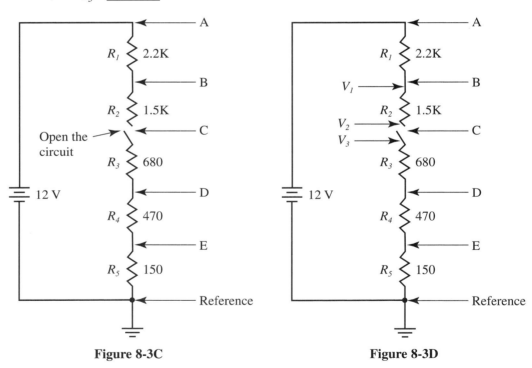

Figure 8-3C   Figure 8-3D

6. How does this test indicate point C is open?

_____

_____

7. Open the circuit at a different point and make the voltage test again.

Summing Up

In your notebook, record any observations, problems, and conclusions for this activity.

Activity 8-4:
Short Circuit Effects

Name _____

Date _____

Class _____ Score _____

Discussion

A short circuit usually causes fuses to blow and smoke to appear within the circuit. This activity is designed to prevent blown fuses while showing the effect of a short circuit.

Objective

In this activity, you will investigate the effects of a short circuit.

Materials and Equipment

One each of the following resistors:
220 Ω
330 Ω
560 Ω
1–Breadboard
1–Power supply
1–VOM
1–Wire jumper

Procedures

1. Construct the circuit shown in **Figure 8-4A.**

2. Connect a wire across R_2 as shown in the detail of **Figure 8-4B.**

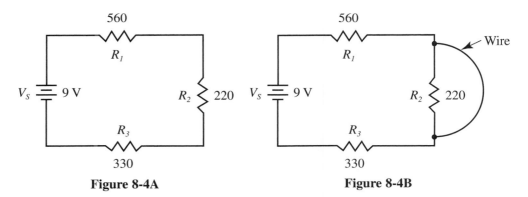

Figure 8-4A   Figure 8-4B

3. Apply power and take a voltage measurement across the resistor.

4. What is the effect of the short across the resistor and why?

_____

_____

_____

Summing Up

In your notebook, record any observations, problems, and conclusions for this activity.

Activity 8-5:
Balanced and Unbalanced
Bridge Circuits

Name _____

Date _____

Class _____ Score _____

Discussion

Although they are simple circuits consisting of only four or five components, bridge circuits are widely used in instrumentation and measurement applications. This activity will demonstrate how the bridge circuit is balanced and indications of an unbalanced bridge circuit.

Objective

In this activity, you will investigate balanced and unbalanced bridge circuits.

Materials and Equipment

1–1K Ω potentiometer (multiturn)
2–1K Ω resistors
1–2.2K Ω resistor
1–4.7K Ω resistor
2–10K Ω resistors
1–Breadboard
1–Power supply
1–VOM

Procedures

Part 1

1. Calculate the currents I_1, I_2, I_3, and I_4 in **Figure 8-5A.**

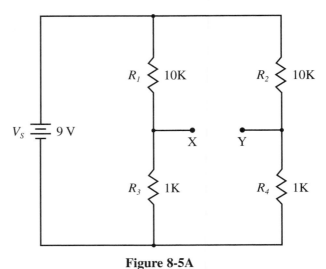

Figure 8-5A

2. Calculate the voltages V_1, V_2, V_3, V_4, and V_{XY}.

3. Record the calculated currents and voltages in the table in **Figure 8-5B.**

(Continued)

	Calculated	Measured
I_1		
I_2		
I_3		
I_4		
V_1		
V_2		
V_3		
V_4		
V_{X-Y}		

Figure 8-5B

4. From the components supply, match as closely as possible the values of the two 1K Ω and two 10K Ω resistors.

5. Using Figure 8-5A, connect the circuit and measure the voltages in step 1. Record these measurements in the table. They should closely compare to the calculated values.

6. Exchange the positions of R_2 and R_4. Calculate and record the values for step 1 in the table in **Figure 8-5C.**

	Calculated	Measured
I_1		
I_2		
I_3		
I_4		
V_1		
V_2		
V_3		
V_4		
V_{X-Y}		

Figure 8-5C

7. Again, measure and record the voltages in step 1. These measurements should closely compare to the calculated values.

8. The voltage V_{XY} may not be exactly zero. Why not?

_____

_____

Part 2

1. Measure the values of the resistors to be used in the circuit shown in **Figure 8-5D.**

2. Record the measured values in the table in **Figure 8-5E.**

(Continued)

Name _____

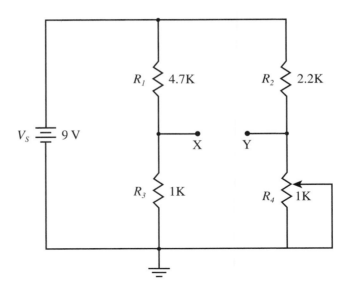

Figure 8-5D

R	Measured
R_1	
R_2	
R_3	

Figure 8-5E

3. With consideration of the measured values, calculate the value of resistance of the potentiometer R_4.

   $R_4 =$ _____

4. Connect the circuit shown in Figure 8-5D.

5. Connect a voltmeter between Points X and Y. Use a scale higher than 9 V.

6. Turn on the power and adjust for the proper source voltage.

7. Slowly adjust the potentiometer until the voltage across bridge points X-Y is 0 V.

8. Switch the meter to a lower range (1 V or 2 V range), and again adjust the potentiometer for 0 V.

9. A closer adjustment can be made by changing the meter to measure a milliamp range and adjusting for 0 mA. With zero current from points X-Y, *this is the balance point of the bridge circuit.*

10. Remove the potentiometer and measure its value. It should be close to the calculated value of step 2.

11. Record the measured value in the table.

Summing Up ——————————————————————————————

In your notebook, record any observations, problems, and conclusions for this activity.

CIRCUIT COMPONENTS 9

Activity 9-1:
General Information about
Circuit Components

Name _____

Date _____

Class _____ Score _____

Discussion

Many types of components do the tasks of electronic circuits. This exercise will provide more review and information on circuit components.

Objective

In this activity, you will become familiar with various circuit components.

Materials and Equipment

1–Pen or pencil

Questions and Problems

1. Why must a switch have the correct current rating?

   _____

2. Draw the common symbol for a fuse.

3. Define *double-pole switch.*

   _____

   _____

(Continued)

4. What do the following switch terms stand for?

   N.O. _____

   N.C. _____

5. What test should be used for fuses?

   _____

6. Why should a miniature lamp be replaced by one with the same identification number?

   _____

   _____

7. Switch ratings are listed for _____ loads.

8. What is the best test for a switch that has no power applied to it?

   _____

9. What are the two types of chemical cells?

   _____

10. What is the proper procedure for the disposal of batteries?

   _____

   _____

Summing Up

In your notebook, record any observations, problems, and conclusions for this activity.

Activity 9-2:
Potentiometer Operation

Name _____

Date _____

Class _____ Score _____

Discussion

A potentiometer, which is a variable resistor, is used to adjust the circuit resistance. As with other types of resistors, a potentiometer has a certain resistance value and power dissipation. Care must be taken not to exceed the maximum power dissipation; otherwise, the resistive element of the potentiometer will be damaged by excessive heat. This may cause the resistive element to open or bring about a change in actual resistance of the potentiometer.

Objective

In this activity, you will investigate the operation of a potentiometer.

Materials and Equipment

1–10K Ω potentiometer
1–Jumper wire (wire with alligator clips on both ends)
1–VOM

Procedures

1. Place the potentiometer at a convenient location on the workbench. A small vise can be used to hold it in position.

2. With the terminals of the potentiometer facing you, turn the shaft of the potentiometer fully clockwise until it stops.

3. Measure the resistance between the terminals, as shown in **Figure 9-2A.**

4. Record the measurement in the clockwise column in the table in **Figure 9-2B.**

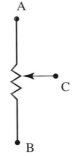

Measure From	Fully Clockwise	Fully Counter-clockwise
A to B		
B to C		
A to C		

Figure 9-2A

Figure 9-2B

5. Turn the shaft of the potentiometer fully counterclockwise.

6. Measure and record the resistance between the terminals as indicated in the counter-clockwise column of the table.

(Continued)

7. Connect the ohmmeter between terminals A and B. Rotate the shaft of the potentiometer and observe the results on the ohmmeter. What is the indication on the ohmmeter?

_____

_____

8. Connect the ohmmeter between terminals A and C. Rotate the shaft of the potentiometer and observe the results on the ohmmeter. What is the indication on the ohmmeter?

_____

_____

9. Connect the ohmmeter between terminals B and C. Rotate the shaft of the potentiometer and observe the results on the ohmmeter. What is the indication on the ohmmeter?

_____

_____

10. Connect a jumper wire between terminals A and C, as shown in **Figure 9-2C.**

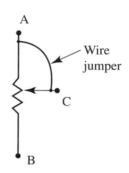

Figure 9-2C

11. Connect the ohmmeter between terminals A and B. Rotate the shaft of the potentiometer and observe the results on the ohmmeter. What is the indication on the ohmmeter?

_____

_____

Summing Up

In your notebook, record any observations, problems, and conclusions for this activity.

NETWORK THEOREMS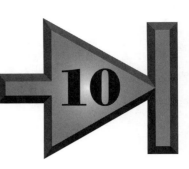

Activity 10-1:
Thevenin's Theorem

Name _____

Date _____

Class _____ Score _____

Discussion

Thevenin's theorem states any resistive circuit, no matter how complex, can be reduced to one resistance in series with a voltage source. This activity will prove Thevenin's theorem using a dc circuit.

Objective

In this activity, you will verify Thevenin's theorem.

Materials and Equipment

One each of the following resistors:
150 Ω
220 Ω
560 Ω
820 Ω
1–Breadboard
1–Dc power supply
1–VOM

Procedures

1. Measure the values of the resistors to be used in **Figure 10-1.**

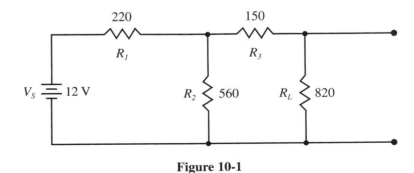

Figure 10-1

(Continued)

2. Using the measured resistor values, calculate the Thevenin equivalent circuit for Figure 10-1.

3. Place the load R_L onto the Thevenin equivalent circuit and calculate V_L and I_L.

4. Connect the circuit shown in Figure 10-1.

5. Measure and record the values of V_L and I_L.

6. Compare the calculated and the measured values. How close are they to the same value?

Summing Up

In your notebook, record any observations, problems, and conclusions for this activity.

Activity 10-2: Superposition Theorem

Name _____

Date _____

Class _____ Score _____

Discussion

The superposition theorem makes it possible to calculate the combined effects of a multisource circuit by observing the individual effects of each source acting alone, then summing those results. Special attention must be paid to *current directions* and *voltage polarities* when applying the superposition theorem. Also, it is important to remove the power supply before replacing it with a short. **Caution! Never short-circuit an operating power supply.**

Objective

In this activity, you will verify the superposition theorem.

Materials and Equipment

One each of the following resistors:
  1.5K Ω
  2.2K Ω
  3.3K Ω
2–9 V power supplies
1–Breadboard
1–VOM

Procedures

1. Measure the actual resistance of each of the resistors.

2. Using the measured resistance values and the superposition theorem, calculate the values in **Figure 10-2A.** Record the calculations in **Figure 10-2B.**

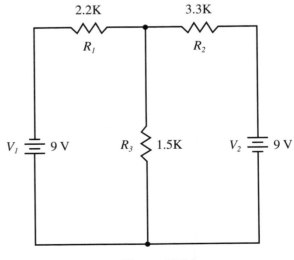

Figure 10-2A

(Continued)

V_1 Active		V_2 Active		V_1 and V_2 Active	
I_{R_1}		I_{R_2}			
V_{R_1}		V_{R_2}		V_1 and V_2 Active	
V_{R_3}		V_{R_3}		V_{R_3}	
I_{R_3}		I_{R_3}		I_{R_3}	

Calculated Data

Figure 10-2B

3. Connect the circuit shown in Figure 10-2A.

4. Replace V_1 with a short and apply power V_2. Measure the values in the table where V_2 is active.

5. Remove the V_1 short.

6. Replace V_2 with a short and apply power V_1.

7. Measure the values in the table where V_1 is active.

8. Remove the V_2 short.

9. Apply power V_1 and V_2. Measure the values of the table in **Figure 10-2C** where V_1 and V_2 are active.

10. How do the calculated and measured values compare?

_____

_____

_____

V_1 Active		V_2 Active		V_1 and V_2 Active	
I_{R_1}		I_{R_2}			
V_{R_1}		V_{R_2}		V_1 and V_2 Active	
V_{R_3}		V_{R_3}		V_{R_3}	
I_{R_3}		I_{R_3}		I_{R_3}	

Measured Data

Figure 10-2C

Summing Up

In your notebook, record any observations, problems, and conclusions for this activity.

ELECTRICAL SOLDERING  11

Activity 11-1:
Tinning a Wire

Name _____

Date _____

Class _____ Score _____

Discussion

Tinning a wire is a routine task in the soldering and assembly of electronic equipment. Often, stranded wire in an electrical assembly must be tinned. This activity provides the necessary information to properly tin a stranded wire.

Objective

In this activity, you will tin the end of a stranded wire.

Materials and Equipment

20-gage stranded wire
Diagonal wire-cutting pliers
Soldering cleaning supplies
Solder pot
Wire stripper

Procedures

1. Cut a piece of stranded wire approximately 5″ long.

2. Carefully strip 1″ of insulation from one end. See **Figure 11-1A.**

   a. Do not strip the insulation in one motion; this will disturb the wire-lay of the strands. Cut the insulation and make a 1/16″ gap.

   b. Remove the insulation with your fingers, using a twisting motion as you pull it off the wire. Twist in the direction of the wire-lay. See **Figure 11-1B.**

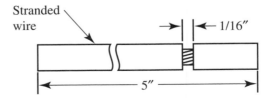

Figure 11-1A

(Continued)

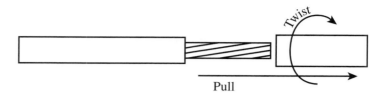

Figure 11-1B

3. Inspect the wire for:

   a. Damage to the wire or insulation.

   b. Disturbed wire-lay.

   c. Clean, even cut of the insulation and wire end.

4. Place a small amount of liquid flux on the end of the wire and dip it into the solder pot. Hold the wire in a vertical position, and slowly lower it into the molten solder. The wire should be held in the solder approximately two seconds. Slowly remove the wire from the solder pot.

5. Clean the tinned wire end with rubbing alcohol.

6. Inspect for:

   a. Exposed copper.

   b. Contour soldering. You should be able to see the wire strands.

   c. Smooth, shiny surface.

   d. Wire or insulation damage.

   e. Wicking of solder under the insulation.

Summing Up

In your notebook, record any observations, problems, and conclusions for this activity.

**Activity 11-2:
Soldering a Wire to an
Eyelet Terminal**

Name _____

Date _____

Class _____ Score _____

Discussion

    Equipment failure, if not the fault of a component, is often the failure of a soldered joint. Such a failure occurs later in the life of the equipment and is the result of someone not taking the time to inspect the work. Other soldering jobs fail inspection because they are not mechanically correct or are poorly soldered. An experienced technician knows that no matter how good a soldered joint looks, it will not pass an inspector's trained eye if it is put together incorrectly. This activity will provide practice in basic soldering.

Objective

In this activity, you will solder a wire to an eyelet terminal.

Materials and Equipment

20-gage stranded wire
60/40 or 63/37 solder
Diagonal wire-cutting pliers
Needle nose pliers
Soldering cleaning supplies
Solder pot and soldering iron
Terminals to be soldered
Wire stripper

Procedures

1. Make a tinned wire.
2. Clean the terminal with a pencil eraser and/or alcohol.
3. Attach the wire to the terminal.
   a. Allow for an insulation clearance of one diameter of the wire insulation. Hold the wire tightly against the back of the terminal. See **Figure 11-2A.**
   b. Pull the end of the wire up and over the terminal. See **Figure 11-2B.**
   c. Cut the end of the wire flush with the top of the terminal. See **Figure 11-2C.**
4. Solder the connection. Place the soldering iron at the back of the terminal, contacting the terminal rather than the wire.
5. Clean the soldered joint for inspection.
6. Turn the soldered sample in to your instructor for inspection and grading. Place your name or other identification on the sample.

(Continued)

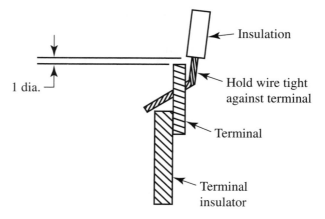

Figure 11-2A

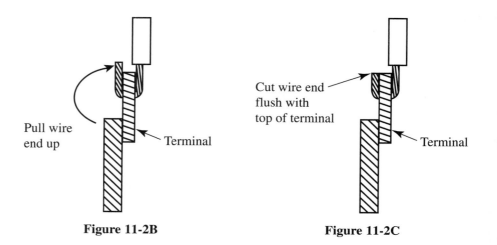

Figure 11-2B   **Figure 11-2C**

Summing Up

In your notebook, record any observations, problems, and conclusions for this activity.

**Activity 11-3:
Soldering Components
on a PC Board**

Name _____

Date _____

Class _____ Score _____

Discussion

A defective soldering job could prevent a CD player from working, cause an incorrect reading on an aircraft fuel gauge, stop a factory assembly line, cause a guided missile to malfunction. When soldering is needed in electronic equipment, it is often required on a printed circuit board assembly. This activity provides practice soldering on printed circuit boards. A printed circuit connection usually requires less solder than a terminal-type connection.

Objective

In this activity, you will solder various components on a printed circuit board.

Materials and Equipment

2–1/4 W resistors
1–1/2 W resistor
3–0.01 µF capacitors, Mylar™ or disc
20-gage stranded wire
22-gage solid wire
60/40 or 63/37 solder
Diagonal wire-cutting pliers
1–PC board, Graymark #69008
Soldering cleaning supplies
Solder pot
Wire stripper
Note: Instructor may substitute other components.

Procedures

1. Using a component lead cleaner, pencil eraser, or rubbing alcohol, clean the component leads and PC board.

2. If needed, tin the leads of the components.

3. Form the leads so they fit into the required holes on the PC board. The components should not have any binding or stress going into the mounting holes.

4. If needed, degrease or clean the PC board and components again.

5. Mount the components onto the PC board as shown in **Figure 11-3A.** This is the component view side of the board. **Figure 11-3B** is the trace side of the PC board. Make sure the components are in the correct holes before soldering.

6. Solder the leads of the components, and cut off the excess lead wire.

7. If needed, clean the PC board, removing all the flux residue.

8. Inspect the PC board for proper component mounting and soldered connections. Record any defects in the table in **Figure 11-3C.**

9. Turn in the PC board and table to your instructor for inspection.

(Continued)

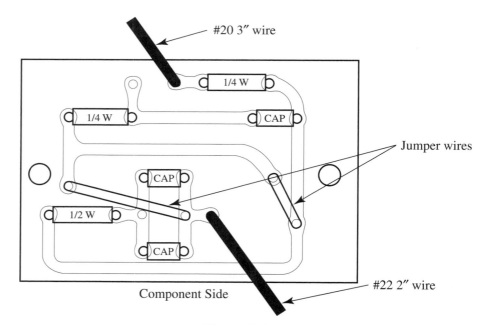

Figure 11-3A

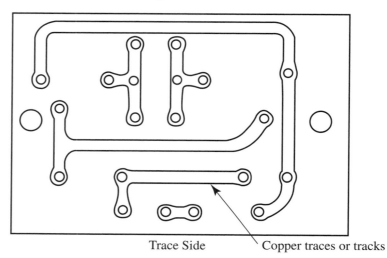

Figure 11-3B

Item	Satisfactory	Defects Found
Resistor mounting		
Capacitor mounting		
Wire mounting		
Soldered connections		

Figure 11-3C

Summing Up

In your notebook, record any observations, problems, and conclusions for this activity.

ASSEMBLY AND REPAIR TECHNIQUES

12

Activity 12-1:
Component-Lead Cleaner

Name _____

Date _____

Class _____ Score _____

Discussion

Before assembling electronic components, oxidation must be removed from component leads to ensure proper soldering. A lead cleaning tool is available from electronic supply companies, but a simple lead cleaner can be made from the braided conductor of a piece of coaxial cable.

Objective

In this activity, you will make a component-lead cleaner.

Materials and Equipment

5″ piece of coaxial cable
Diagonal wire-cutting pliers
File
Lead free solder
Soldering iron
Wire stripper

Procedures

1. Remove 3 1/2″ of the outer insulation from a piece of coaxial cable. See **Figure 12-1A.**

2. Remove the outer braid from the center conductor and insulation.

3. Push the braid in on itself and flatten it.

4. Using lead free solder, solder 1/4″ of the braid ends. See **Figure 12-1B.**

5. Trim and file any braid wire ends that may extend beyond the soldered area. See **Figure 12-1C.**

6. Fold the braid in the center to make a completed lead cleaner.

(Continued)

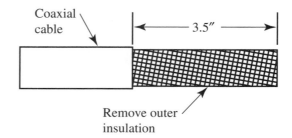

Coaxial cable

3.5″

Remove outer insulation

Figure 12-1A

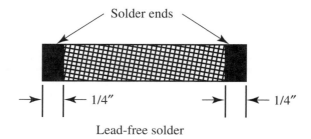

Solder ends

1/4″   1/4″

Lead-free solder

Figure 12-1B

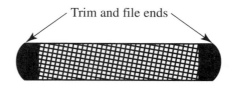

Trim and file ends

Figure 12-1C

Summing Up

In your notebook, record any observations, problems, and conclusions for this activity.

Activity 12-2: Multiconnector

Name _____

Date _____

Class _____ Score _____

Discussion

In the assembly of electronic equipment, a cable is often required for connection to another piece of equipment. It would be convenient to purchase ready-made cables, but often they are not wired correctly for the needs of the equipment. Many times a cable must be specially made to match the electrical connections of the equipment. In this activity you have an opportunity to sharpen your skills by soldering in the confined area of a D-connector.

Objective

In this activity, you will solder wires into a D-type connector.

Materials and Equipment

22-gage stranded wire
Assembly instructions (see instructor)
D-connector, Radio Shack #276-1537c
Diagonal wire-cutting pliers
Needle nose pliers
Soldering cleaning supplies
Soldering iron and solder, 63/37 preferred
Solder pot
Solder wick, size 1
Wire stripper

Procedures

1. Following the instructions given by your instructor, solder the wires into the D-connector. See **Figures 12-2A** and **12-2B**.

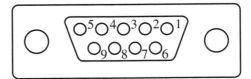

Figure 12-2A

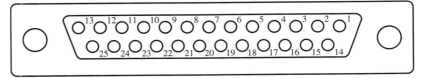

Figure 12-2B

(Continued)

Summing Up ———————————————————————

In your notebook, record any observations, problems, and conclusions for this activity.

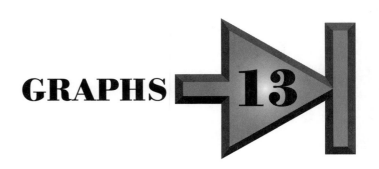

GRAPHS 13

Activity 13-1:	Name _____
Making a Dc Circuit Graph	Date _____
	Class _____ Score _____

Discussion

Graphs visually show electronic circuit action. They often show the strength of the voltage or current in relationship to time. Although the current starts flowing immediately when a switch is closed, the current does not get to maximum for a period of time. Only a graph can show this relationship.

Objective

In this activity, you will graph the current of a resistive dc circuit.

Materials and Equipment

1–Calculator (optional)
1–Pen or pencil
1–Ruler (optional)

Procedures

1. Calculate the total maximum current for **Figure 13-1A.**

2. Write the scale in amps on the Y-axis of the graph in **Figure 13-1B.**

3. Write the scale in milliseconds on the X-axis of the graph.

4. Assuming it takes 6 ms for the current to reach maximum, plot the current of Figure 13-1A on the graph.

Summing Up

In your notebook, record any observations, problems, and conclusions for this activity.

(Continued)

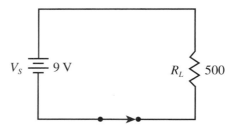

Figure 13-1A

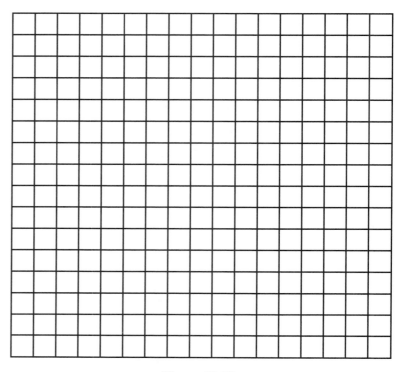

Figure 13-1B

**Activity 13-2:
Making a Dc Linear Graph**

Name _____

Date _____

Class _____ Score _____

Discussion

This activity shows the linear relationship of the voltage and current in a dc circuit. Depending on the equation used in calculating the values, line graphs can be linear or non-linear. When the voltage changes, the current changes by the same amount; therefore, the graph is linear.

Objective

In this activity, you will graph the current of a resistive dc circuit under changing voltage.

Materials and Equipment

1–Breadboard
1–Calculator (optional)
1–Pen or pencil
1–Ruler (optional)
1–VOM

Procedures

1. Calculate the current in **Figure 13-2A** for each voltage indicated in the table in **Figure 13-2B.** Record the answers in milliamps.

2. Plot the current information on graph paper.

3. Connect the circuit shown in Figure 13-2A.

4. Measure the current for each of the required voltages.

5. Plot the measured current on the same graph paper as in step 2.

6. What identifies this graph as being a linear graph?

   _____

   _____

7. Why is there a difference between the calculated and measured values?

   _____

   _____

Summing Up

In your notebook, record any observations, problems, and conclusions for this activity.

(Continued)

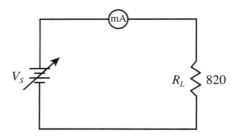

Figure 13-2A

V_S	I (Calculated)	I (Measured)
1		
2		
3		
4		
5		
6		
7		
8		
9		
10		

Figure 13-2B

Activity 13-3: Making a Dc Nonlinear Graph

Name _____

Date _____

Class _____ Score _____

Discussion _____

This activity shows the nonlinear relationship of the voltage and power of a dc circuit due to the equation $P = \dfrac{E^2}{R}$.

Objective _____

In this activity, you will graph the power of a resistive dc circuit under changing voltage.

Materials and Equipment _____

1–Calculator (optional)
1–Pen or pencil
1–Ruler (optional)

Procedures _____

1. Calculate the power in **Figure 13-3A** for each voltage indicated in the table in **Figure 13-3B.** Record the answers in milliwatts.

2. Plot the information on graph paper.

3. What identifies this graph as being a nonlinear graph?

   _____

   _____

   _____

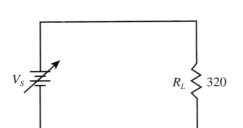

Volts	P (Calculated)
3	
6	
9	
12	
15	
18	
21	

Figure 13-3A

Figure 13-3B

(Continued)

Summing Up _____

In your notebook, record any observations, problems, and conclusions for this activity.

TRIGONOMETRY FOR ELECTRICITY

Activity 14-1:
Trigonometric Applications

Name _____

Date _____

Class _____ Score _____

Discussion

Trigonometry is the study of relationships between the sides and angles of triangles. It is used extensively in alternating current; however, trigonometry is applied to other areas as well. This activity will give examples of other applications.

Objective

In this activity, you will practice two applications of trigonometry.

Materials and Equipment

1–Calculator
1–Pen or pencil

Procedures

1. Use a calculator to find the angle of each of the following trigonometric functions.

   a. sin = 0.42262 = _____

   b. cos = 0.871557 = _____

   c. tan = 1.19175= _____

   d. sin = 0.6018 = _____

   e. cot = 0.42447 = _____

   f. cos = 0.39073 = _____

   g. tan = 0.08749 = _____

   h. cot = 2.74747 = _____

2. In **Figure 14-1A,** the only guy wire available will make three 55′ pieces. If the pole is 45′ high, how far out from the bottom of the pole will the guy wire anchors be placed?

(Continued)

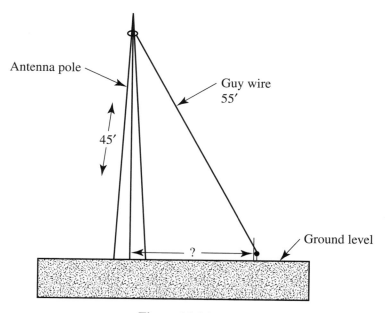

Figure 14-1A

3. Your task is to measure the field strength of a transmitter. You are required to know the distance to the transmitter from the point of measurement. You can see the transmitter tower from where you stand, **Figure 14-1B.** Unfortunately, a ravine, stream, and forest prevent you from measuring the distance directly to the transmitter tower.

A water tower is visible straight down the highway. Using a surveying transit and placing the water tower at 0°, you measure the transmitter tower at 43.5°. You measure down the highway and find it is 1200′ to the water tower. From the point of the field strength measurement, what is the distance to the transmitter?

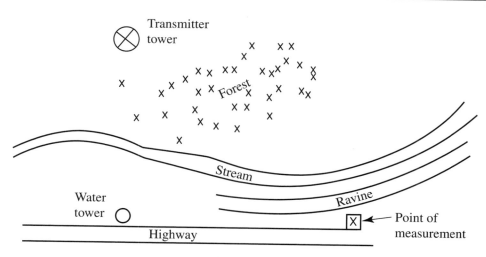

Figure 14-1B

Summing Up

In your notebook, record any observations, problems, and conclusions for this activity.

MAGNETISM

Activity 15-1:
Action of an Electromagnet

Discussion

Electromagnets are used in such small devices as relays, water or natural gas valves, doorbells, automobile power door locks, and mechanical assemblies in audio/video tape players. The strength of the magnetic field of an electromagnet is increased by more turns or greater current.

Objective

In this activity, you will demonstrate the action of an electromagnet.

Materials and Equipment

7′ of #26 or #30 magnet wire
1–10 or 12 penny nail
1–Aluminum nail
1–Battery (D-cell)
Masking tape

Procedures

1. Cut a 7′ length of #26 or #30 magnet wire, and remove 1/2″ insulation from each end.

2. Wind as many turns of magnet wire on the nail as possible, taking care not to cross over any turns. See **Figure 15-1.** If the turns cross over each other, the generated magnetic field will be smaller than it should be. Insulated hookup wire may be used, but the resulting magnetic field will not be as strong. Secure the wire with masking tape.

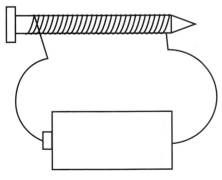

Figure 15-1

(Continued)

3. Connect the ends of the magnet wire to the battery or other power source. Allow for one wire end to be easily disconnected or insert a switch in the circuit.
Caution! If a higher voltage is used, a higher current can produce a stronger magnetic field, but it can also cause the wire to become hot. The insulation will burn off the wire and cause a short circuit across the voltage source.

4. Bring the electromagnet close to a paper clip. What do you observe?

_____

_____

5. Use an aluminum nail or other small aluminum object. What do you observe and why?

_____

_____

_____

_____

_____

Summing Up

In your notebook, record any observations, problems, and conclusions for this activity.

Activity 15-2:
Magnetic Induction

Name _____

Date _____

Class _____ Score _____

Discussion

Magnetic fields can pass through materials without any actual physical contact between them. The magnetic field of a magnet penetrates the other object, and the object becomes part of the magnetic circuit. Any other object brought close to the magnetic field also becomes part of the magnetic circuit. How well the magnetic field goes through the other objects depends on their material makeup.

Objective

In this activity, you will demonstrate the action of magnetic induction.

Materials and Equipment

1–Bar magnet
2–Metal paper clips
1–Small nail

Procedures

1. Use the bar magnet to pick up the paper clips and nail to verify they are magnetic.

2. Allow the bar magnet to come close but not touch the nail. See **Figure 15-2A.**

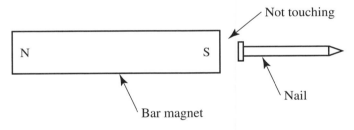

Figure 15-2A

3. While moving the magnet and nail at the same time, bring the end of the nail close to one of the paper clips. See **Figure 15-2B.** What happens to the paper clip?

_____

_____

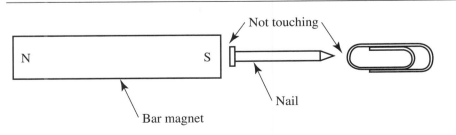

Figure 15-2B

(Continued)

4. Bring the end of the paper clip close to the second paper clip. What do you observe and why?

_____

_____

_____

_____

_____

Summing Up

In your notebook, record any observations, problems, and conclusions for this activity.

**Activity 15-3:
Magnetizing and
Demagnetizing Objects**

Name _____

Date _____

Class _____ Score _____

Discussion

All magnetic materials retain some magnetic flux after exposure to a magnetic field. The magnetism that remains in an object is called residual magnetism. Permanent magnets have a high retentivity or a high residual magnetism. The residual magnetism can be used in such items as a screwdriver to help hold a screw on the end until the screw starts to thread. At other times, it may be undesirable to have a screwdriver magnetized. A coil operating with alternating current will demagnetize it.

Objective

In this activity, you will demonstrate the action of magnetizing and demagnetizing objects.

Materials and Equipment

1–Small metal paper clip
1–Small metal screwdriver (nonmagnetized tip)
1–Soldering gun

Procedures

Warning! The soldering gun tip is extremely hot.

1. Place a small screwdriver inside the tip of the soldering gun. See **Figure 15-3A.** Avoid touching the tip to the screwdriver.

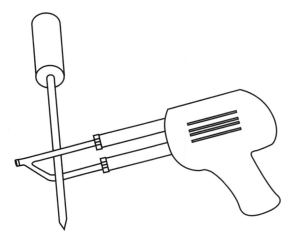

Figure 15-3A

2. Turn the soldering gun on and slowly pull the screwdriver out of the tip. See **Figure 15-3B.**

(Continued)

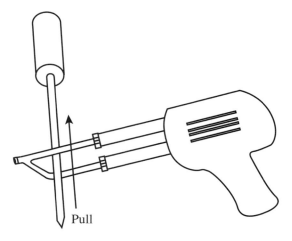

Pull

Figure 15-3B

3. Try picking up the paper clip with the end of the screwdriver. What happens to the paper clip? Is the screwdriver still magnetized?

_____

_____

_____

4. Place the end of the screwdriver back into the soldering tip.

5. Turn the soldering gun on. Be careful not to move the screwdriver.

6. Turn the soldering gun off.

7. Try picking up the paper clip with the end of the screwdriver. What happens to the paper clip? Is the screwdriver magnetized?

_____

_____

_____

8. How do you explain the action of the soldering gun on the screwdriver?

_____

_____

_____

_____

Summing Up

In your notebook, record any observations, problems, and conclusions for this activity.

ALTERNATING CURRENT ▶ 16 ◀

Activity 16-1:
Ac Measurement
Relationships

Name _____

Date _____

Class _____ Score _____

Discussion

This activity compares the measurements of dc and ac voltages around a series circuit. To eliminate meter insertion error, the current is calculated using Ohm's law with the applied voltage and measured total resistance.

Objective

In this activity, you will investigate the relationship of dc, RMS, and peak-to-peak voltage measurements.

Materials and Equipment

One each of the following resistors:
- 150 Ω
- 220 Ω
- 470 Ω
- 1–Ac power supply
- 1–Breadboard
- 1–Dc power supply
- 1–VOM

Procedures

Part 1: Dc

1. Using a dc source, connect the circuit as shown in **Figure 16-1A.**

2. Measure the total resistance and record it in the upper left-hand corner of the table in **Figure 16-1B.**

3. Using the measured resistance, calculate the current and record it in the table.

4. Using 12 V as the source, calculate voltages V_1, V_2, V_3 and record them in the calculated value column of the table.

5. Turn on the power. Adjust the voltage source for 12 V.

6. Measure voltage drops V_1, V_2, V_3 and record them in the dc column of the table.

7. Calculate the power ($I_T \times V_T$) and record it in the dc column of the table.

(Continued)

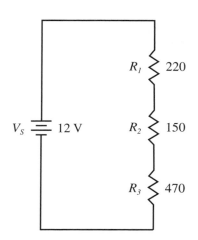

Figure 16-1A

R_T mea. = ___	Calculated Value	Dc	Ac-RMS	Ac P-P
Applied Voltage	– – – –	12	12	34
I_T				
V_1				
V_2				
V_3				
Calculated Power				

Figure 16-1B

Part 2: Ac RMS

1. Replace the dc source with a variable ac source.

2. Calculate the current and record it in the Ac-RMS column of the table.

3. Adjust the voltage source for 12 V and turn on the power.

4. Measure voltage drops V_1, V_2, V_3 and record them in the Ac-RMS column of the table.

5. Calculate the power ($I_T \times V_T$) and record it in the Ac-RMS column of the table.

Part 3: Ac Peak-to-Peak

Note: Ask your instructor for assistance with this part of the activity.

1. Adjust the ac source to 34 V_{P-P}.

2. Calculate the current and record it under the Ac P-P column of the table.

3. Measure voltage drops V_1, V_2, V_3 and record them in the Ac P-P column of the table.

4. Calculate the power ($I_T \times V_T$) and record it in the P-P column of the table.

Part 4: Questions

1. How are the three measurements the same or different?

_____

_____

_____

_____

_____

2. Why are the peak-to-peak measurements *not* different?

_____

_____

_____

Summing Up ————

In your notebook, record any observations, problems, and conclusions for this activity.

THE OSCILLOSCOPE

Activity 17-1:
Oscilloscope Controls

Name _____

Date _____

Class _____ Score _____

Discussion

At times an oscilloscope displays a waveform that cannot be properly seen. An inexperienced technician may search for the correct knob by turning first one and then another. Quite by luck, the oscilloscope may be adjusted to make the display behave properly. A technician who understands the oscilloscope knows exactly which knob will cause the waveform to be correctly displayed without resorting to trial and error.

Objective

In this activity, you will investigate the controls of an oscilloscope.

Materials and Equipment

1–100K potentiometer
1–Audio signal generator
1–Breadboard
1–Oscilloscope

Procedures

1. Set up the equipment as shown in **Figure 17-1A.**

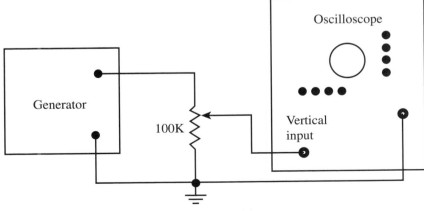

Figure 17-1A

(Continued)

2. Turn on the power, and observe the reaction on the oscilloscope display when each of the controls in the table in **Figure 17-1B** is adjusted. Record any notes about each control.

3. Demonstrate and explain each control adjustment to your instructor.

✓	Control	Notes
	Intensity	
	Focus	
	Vertical Position	
	Horizontal Position	
	Vertical Attenuator	
	Horizontal Time Base	
	Vertical Input Switch	
	Trigger Level	

Figure 17-1B

Summing Up

In your notebook, record any observations, problems, and conclusions for this activity.

Activity 17-2:	Name _____
Measuring Peak-to-Peak	Date _____
Voltage	Class _____ Score _____

Discussion

Measuring peak-to-peak voltage with an oscilloscope is sometimes necessary when troubleshooting. Also, many laboratory experiments require the input signal be adjusted to a certain peak-to-peak value for proper operation.

Objective

In this activity, you will adjust and read peak-to-peak measurements using an oscilloscope.

Materials and Equipment

1–100K potentiometer
1–Audio signal generator
1–Breadboard
1–Oscilloscope

Procedures

1. Set up the equipment as shown in **Figure 17-2A.**

2. Check the oscilloscope calibration.

3. Adjust the generator for the required peak-to-peak voltages in the table in **Figure 17-2B.**

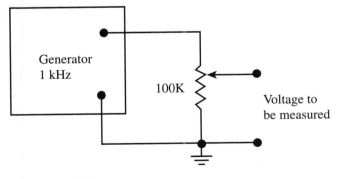

Figure 17-2A

Voltage	Instructor Check
6.0	
2.8	
8.4	
1.0	
0.25	
3.0	
0.015	

Figure 17-2B

Summing Up

In your notebook, record any observations, problems, and conclusions for this activity.

**Activity 17-3:
Comparing Measurements**

Name _____

Date _____

Class _____ Score _____

Discussion

An oscilloscope measures the peak-to-peak value of a signal. This activity compares the measurements made with an oscilloscope to those found when using a VOM.

Objective

In this activity, you will compare peak-to-peak measurement with a VOM reading.

Materials and Equipment

1–100K potentiometer
1–Audio signal generator
1–Breadboard
1–Oscilloscope
1–VOM

Procedures

1. Set up the equipment as shown in Figure 17-2A of Activity 17-2.

2. Check the oscilloscope calibration.

3. Adjust the generator for maximum output.

4. Adjust the potentiometer for a peak-to-peak voltage that is easy to read. Measure and record the peak-to-peak voltage in the table in **Figure 17-3.**

5. Measure the output voltage (ac) with a VOM, and record the result in the table.

6. Why is there a difference in the measurements?

_____

_____

Instrument	Voltage
Oscilloscope	
Voltmeter	

Figure 17-3

Summing Up

In your notebook, record any observations, problems, and conclusions for this activity.

Activity 17-4:
Measuring Dc Voltages

Name _____

Date _____

Class _____ Score _____

Discussion

An experienced technician with an oscilloscope has no need for an ordinary dc voltmeter. Technicians can rely on the oscilloscope to make any measurement needed, even very small dc voltages.

When dc voltage is measured, the trace of the oscilloscope moves upward. However, if the power source is connected with the polarity reversed, the oscilloscope will deflect downward. In this case, the reference is placed at the top of the oscilloscope screen rather than at the bottom. In this way, both positive and negative dc voltage can be measured with an oscilloscope.

The circuit in this activity is a simple series resistive circuit that will be measured as a voltage divider. Remember, this is an oscilloscope activity, not a voltage divider activity. The objective is to learn to measure dc voltages using an oscilloscope.

Caution! The ground of the oscilloscope must be connected to the circuit ground only. Any attempt to measure the voltage drops across the resistors may result in overheated resistors or a damaged oscilloscope probe.

Objective

In this activity, you will measure dc voltages using an oscilloscope.

Materials and Equipment

One each of the following resistors:
    680 Ω
    1.5K Ω
    4.7K Ω
2–150 Ω resistors
1–Breadboard
1–Dc power supply
1–Oscilloscope

Procedures

1. Connect the circuit as shown in **Figure 17-4A.**

2. Connect the oscilloscope probe ground to the circuit ground.

3. Move the vertical input switch to the ground position, and adjust the trace for reference at the bottom of the oscilloscope screen. See **Figure 17-4B.**

4. Connect the oscilloscope ×1 probe to point A.

5. Move the vertical input switch to the dc position.

6. Adjust the VOLTS/DIVISION control to 2 V per division.

7. Adjust the voltage so the trace moves up on the graticule six divisions.

   Number of divisions × Volts/Division = _____ V

   6 divisions × 2 V/Division = 12 V

8. Record the results in the table in **Figure 17-4C.**

(Continued)

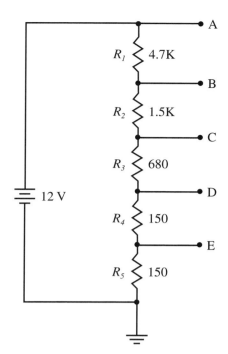

Figure 17-4A

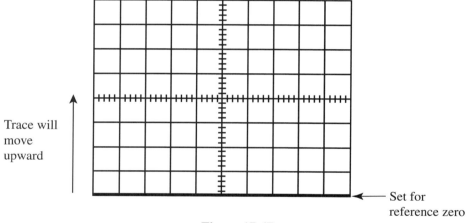

Trace will
move
upward

Set for
reference zero

Figure 17-4B

Point	Voltage
A	
B	
C	
D	
E	

Figure 17-4C

9. Connect the oscilloscope probe to point B.

10. Adjust the vertical VOLTS/DIVISION control until the trace is at the maximum height on the graticule.

11. Read the voltage.

    Number of divisions × Volts/Division = _____ V

12. Repeat steps 10 and 11 for the other points in the diagram, and record the results in the table.

    The voltages of points A through E are found to be 12, 4.1, 1.64, 0.5, and 0.25 using Ohm's law. The voltages recorded in the table should be close to these. They will not be exactly the same because of variations in resistor tolerances and oscilloscope calibrations. For more practice, substitute other resistors and make the measurements again.

Summing Up

In your notebook, record any observations, problems, and conclusions for this activity.

WAVEFORMS AND MEASUREMENTS

18

Activity 18-1:
Measuring Percentage
of Tilt

Name _____

Date _____

Class _____ Score _____

Discussion

Tilt is a square wave distortion. It can either be desirable or undesirable, depending on the application. A square wave can be used to test an amplifier. The amount of tilt can give an indication as to its high-frequency operation. In this activity, a generator supplies a square wave to a resistor-capacitor network which distorts the square wave and provides tilt for practice in measurements.

Objective

In this activity, you will measure the tilt percentage of a square wave.

Materials and Equipment

One each of the following resistors:
   1K Ω
   1.5K Ω
   2.2K Ω
   2.7K Ω
   5.6K Ω
2–0.1 µF capacitor
1–Audio generator (square wave)
1–Oscilloscope

Procedures

1. Using a 1K Ω resistor, build the circuit shown in **Figure 18-1A.**

2. Adjust the generator for a square wave output at 10 kHz. Adjust the output amplitude and oscilloscope for a readable waveform on the oscilloscope.

3. Measure the A and B measurements and record them in the table in **Figure 18-1B.**

4. Replace the 1K Ω resistor with each of the resistors listed in the table. Record the A and B measurements.

5. Calculate the percentage of tilt for each resistance value.

(Continued)

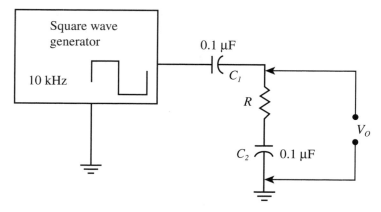

Figure 18-1A

R	A	B	Tilt Percentage
1K			
1.5K			
2.2K			
2.7K			
5.6K			

Figure 18-1B

6. How does the resistance affect the amount of tilt?

_____

_____

7. What change in tilt occurs if you change the value of the capacitance?

_____

_____

Summing Up

In your notebook, record any observations, problems, and conclusions for this activity.

Activity 18-2:	Name _____
Measuring Percentage	Date _____
of Duty Cycle	Class _____ Score _____

Discussion

Duty cycle, or the ratio of the pulse width to the time period of the wave, is an important term used in the discussion of square waveforms. Since a pure square wave always has equal positive and negative alternation, the duty cycle will always be 50%. For a pulse waveform, the duty cycle is something less than 50% and indicates the pulse duration is a certain percentage of the total time period.

Objective

In this activity, you will measure the percentage of duty cycle.

Materials and Equipment

1–0.01 µF capacitor
1–0.1 µF capacitor
1–Oscilloscope
1–Unknown circuit generator (see instructor)

Procedures

1. Ask your instructor for an unknown circuit for measuring the duty cycle.

2. Connect the oscilloscope probe to the test point.

3. Turn on the power.

4. Measure the time the pulse is high and the time period of the waveform. Record these in the table in **Figure 18-2.**

C_2	Time High	Time Period	Duty Cycle
0.01 µF			
0.1 µF			

Figure 18-2

5. Change capacitor C_2 to 0.1 µF.

6. Measure the time the pulse is high and the time period of the waveform. Record these in the table.

7. Calculate the duty cycle percentage. How do the two duty cycles compare?

_____

_____

(Continued)

8. What is the frequency of each of the waveforms?

f_1 = _____

f_2 = _____

Summing Up

In your notebook, record any observations, problems, and conclusions for this activity.

Activity 18-3:
Measuring Phase Angle

Name _____

Date _____

Class _____ Score _____

Discussion

In this activity, you will use both the dual trace and Lissajous methods to measure phase angles, then compare the two methods. This will provide practice in phase measurement.

Objective

In this activity, you will practice measuring phase angle using an oscilloscope.

Materials and Equipment

1–120:12.6 V ac filament transformer
1–10K potentiometer
1–1 µF capacitor
1–Audio generator
1–Breadboard
1–Knob for the potentiometer (It should have a mark to indicate shaft position.)
1–Oscilloscope
3″ × 5″ index card, cut in half and taped under the knob of the potentiometer to mark shaft positions

Procedures

Part 1: Dual Trace Measurement Method

1. Connect the variable phase shift circuit of **Figure 18-3A.**

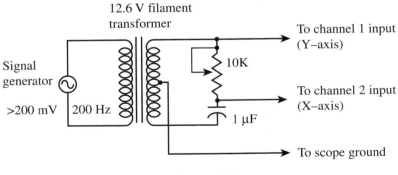

Figure 18-3A

2. Connect the oscilloscope to the circuit as indicated in the circuit figure.

3. Adjust the generator output for greater than 200 mV at 200 Hz.

4. Adjust the oscilloscope controls to achieve superimposed waveforms similar to those in **Figure 18-3B.** You may need to adjust the potentiometer slightly to get a phase difference.

(Continued)

5. Mark the position of the potentiometer.

6. Measure the phase angle and record it in the table in **Figure 18-3C.**

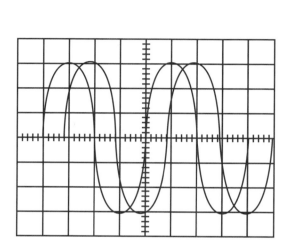

Mark	Method	
	Dual	Lissajous
	Angle	Angle
1		
2		
3		
4		
5		
6		
7		
8		

Figure 18-3B   **Figure 18-3C**

7. Turn the shaft of the potentiometer approximately 1/8 turn, mark its position, and measure the phase angle. Record it in the table.

8. As in Step 6, keep turning the potentiometer 1/8 turns, measuring and recording the phase angles. Make sure each position is marked.

Part 2: Lissajous Measurement Method

1. Return the shaft of the potentiometer to the position in Step 4 of Part 1.

2. Change the oscilloscope to measure phase using Lissajous figures.

3. Measure the phase angle and record it in the table.

4. Turn the shaft of the potentiometer to the next mark, measure the phase, and record it in the table.

5. Keep turning the potentiometer to each previous mark, measuring and recording the phase angle.

Part 3: Questions

1. How do the dual trace and Lissajous methods compare in their measurements?

_____

_____

_____

2. Which method do you feel is the most accurate and why?

_____

_____

_____

Summing Up

In your notebook, record any observations, problems, and conclusions for this activity.

RESISTIVE AC CIRCUITS  19

Activity 19-1:
Phase in Ac Resistive
Circuits

Name _____

Date _____

Class _____ Score _____

Discussion

In a purely ac resistive circuit, the voltage and current are in step (in phase) with each other. Other components, such as inductors and capacitors, cause the voltage and current to be out of phase with each other. This activity gives you an opportunity to observe sine waves that are in phase.

Objective

In this activity, you will measure the phase angle of a resistive ac circuit.

Materials and Equipment

One each of the following resistors:
  220 Ω
  820 Ω
1–Audio generator
1–Breadboard
1–Oscilloscope

Procedures

1. Connect the circuit shown in **Figure 19-1A.**

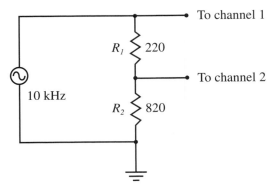

Figure 19-1A

(Continued)

2. Adjust the generator for 10 kHz sine wave output at maximum amplitude.

3. Using a dual trace oscilloscope, observe the phase of the voltage drop across R_2 with reference to the source voltage.

4. On **Figure 19-1B,** draw the waveforms observed.

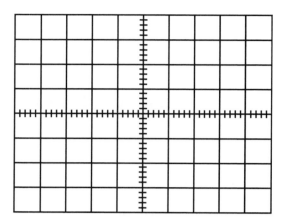

Figure 19-1B

5. What is the phase difference between the source voltage and the voltage across R_2?

_____

_____

Summing Up

In your notebook, record any observations, problems, and conclusions for this activity.

INACTANCE

Activity 20-1:
Inductive Reactance Effects

Name _____

Date _____

Class _____ Score _____

Discussion

Due to the storage of magnetic energy, an inductor has a reactance toward alternating current. The storage of magnetic energy and the amount of reactance depends on both the inductance and frequency of the alternating current. As the reactance increases, the series current decreases, causing the voltage drops to decrease.

Objective

In this activity, you will investigate the effects of inductive reactance.

Materials and Equipment

1–10K Ω resistor
1–Audio generator
1–Breadboard
1–Inductor (see instructor)
1–Oscilloscope

Procedures

1. Connect the circuit as shown in **Figure 20-1A.**

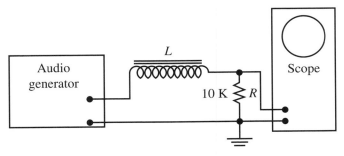

Figure 20-1A

2. Set the generator for maximum output voltage.

(Continued)

3. Set the generator for 200 Hz and measure the peak-to-peak voltage. Record this measurement in the table in **Figure 20-1B.** This input voltage must be maintained for each of the other frequencies in the table.

Frequency	Volts P-P
500 Hz	
1 kHz	
2 kHz	
5 kHz	
10 kHz	
15 kHz	
20 kHz	

Figure 20-1B

4. Adjust the generator to each of the other frequencies indicated in the table, and record the peak-to-peak voltages.

5. Draw a frequency-to-voltage graph.

6. At what frequency does the voltage start to decrease? Why? _____

_____

_____

Summing Up

In your notebook, record any observations, problems, and conclusions for this activity.

	Activity 20-2:	Name
	Testing Inductors	Date
		Class _____ Score _____

Discussion

Testing an inductor consists of checking for continuity. If the inductor has continuity, it is usually good. Always remember to disconnect the inductor from the circuit when checking for resistance or continuity. This will prevent any parallel circuit paths from entering into the meter results. An inductance bridge can be used to test an inductor; however, the inductance will seldom change and is most critical at high frequencies.

Objective

In this activity, you will test whether an inductor is good, shorted, or open.

Materials and Equipment

1–Inductance bridge (optional)
4–Inductors (see instructor)
1–VOM

Procedures

1. Using an ohmmeter, test each inductor for continuity and record the results in the table in **Figure 20-2.**

2. Check the inductance of each of the inductors and record the inductance in the table.

Inductor	Good	Open	Shorted	Inductance
1				
2				
3				
4				
5				
6				

Figure 20-2

Summing Up

In your notebook, record any observations, problems, and conclusions for this activity.

Activity 20-3:
Inductive Reactance in Series

Name _____

Date _____

Class _____ Score _____

Discussion

Although the total inductance in series is the sum of the inductances, the total inductive reactance is also added. This presents a total reactance toward an alternating current. Remember Ohm's law still applies to alternating current as do the series and parallel circuit rules. An oscilloscope must be used because of the high frequency of the source voltage.

Objective

In this activity, you will investigate the effects of inductive reactance in series.

Materials and Equipment

2–2.5 mH inductors (see instructor)
1–820 Ω resistor
1–Audio generator (capable of 100 kHz)
1–Breadboard
1–Jumper wire
1–Oscilloscope

Procedures

1. Connect the circuit as shown in **Figure 20-3A.**

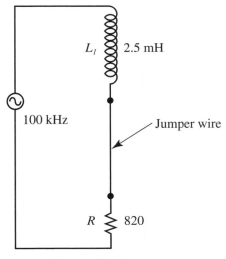

Figure 20-3A

2. Set the generator for maximum output voltage.

3. Set the generator for 100 kHz, and measure the peak-to-peak voltage across the resistor.

$V_{P-P} =$ _____

(Continued)

4. Remove the jumper wire and insert the second inductor in place of the wire. See **Figure 20-3B.**

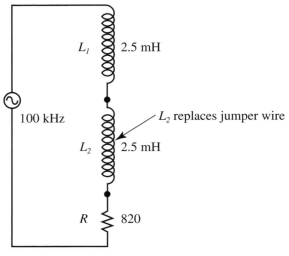

L_1 2.5 mH

100 kHz

L_2 replaces jumper wire

L_2 2.5 mH

R 820

Figure 20-3B

5. Measure the peak-to-peak voltage again.

   $V_{P-P} = $ _____

6. Why is the voltage smaller in step 5?

   _____

   _____

Summing Up

In your notebook, record any observations, problems, and conclusions for this activity.

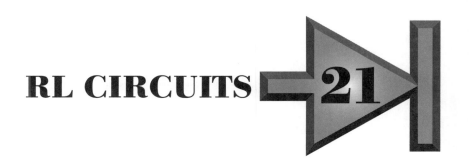

RL CIRCUITS — 21

Activity 21-1:
RL Circuit Operation

Name _____

Date _____

Class _____ Score _____

Discussion

Resistance changes in an RL circuit cause the impedance and the voltage-current phase angle to change. Because the voltage and current in a resistor are in phase, the voltage across a resistor is representive of its current. In this activity, the source voltage is compared to the voltage across the resistance (current flowing in the resistor).

Objective

In this activity, you will investigate the effects of resistance in an inductive circuit and the related phase angle.

Materials and Equipment

One each of the following resistors:
270 Ω
680 Ω
1500 Ω
4700 Ω
10K Ω
1–2.5 mH inductor (see instructor)
1–Audio generator
1–Breadboard
1–Oscilloscope

Procedures

1. Connect the circuit as shown in **Figure 21-1A** with the 270 Ω resistor.

2. Set the generator for 3/4 maximum output voltage.

3. Set the generator for 100 kHz.

4. Connect channel 1 of the oscilloscope on the source side of the inductor and channel 2 at the top side of the resistor.

5. Using a method for good accuracy, adjust the oscilloscope controls to measure phase angle.

6. Measure the phase angle of the two voltages, and record them in the table in **Figure 21-1B.**

(Continued)

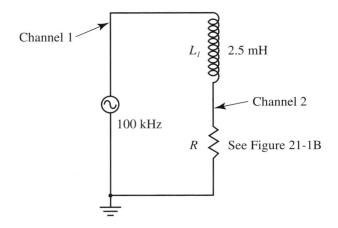

Figure 21-1A

R	Divisions out of Phase	Degrees out of Phase	Calculated Phase Angle
270			
680			
1500			
4700			
10K			

Figure 21-1B

7. Replace the resistor with the next resistor listed in the table, and measure the phase angle again.

8. Repeat for the remaining resistors.

9. Calculate the phase angle for each of the resistances in the table.

10. How do the measured and calculated phase angles compare?

_____

_____

_____

11. On a separate sheet of paper, draw a vector diagram for each of the phase conditions listed in the table. (Optional)

Summing Up ———————————————————————————

In your notebook, record any observations, problems, and conclusions for this activity.

TRANSFORMERS 22

**Activity 22-1:
Transformer Voltage and
Turns Ratio**

Name _____

Date _____

Class _____ Score_____

Discussion

Changing transformer voltages from primary to secondary depends on the ratio of the number of turns in the primary to number of turns in the secondary. The turns ratio is an important factor in the structure of a transformer and is sometimes specified in a replacement transformer. Since the turns ratio and the voltage ratio are equal, the voltage ratio can be used to calculate the turns ratio.

Objective

In this activity, you will measure the voltage and turns ratio of a transformer.

Materials and Equipment

1–16A3 transformer
1–16A10 transformer
1–Audio generator
1–Breadboard
1–Oscilloscope

Procedures

1. Construct the circuit shown in **Figure 22-1** using a 16A10 transformer.

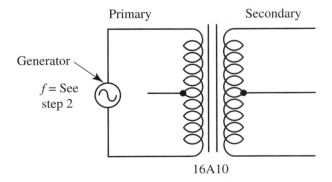

Figure 22-1

(Continued)

2. Adjust the generator for the operating frequency of the transformer and the output amplitude for a readable waveform on the oscilloscope. An output voltage of 1 V or 0.1 V will make calculations easier.

3. Measure the input and output peak-to-peak voltages.

   $V_{IN} = $ _____

   $V_{OUT} = $ _____

4. Calculate the turns ratio of the transformer. (Reduce the fraction.)

   Turns ratio $= \dfrac{V_{IN}}{V_{OUT}}$

   $= $ _____ : _____

5. Using the preceding steps, find the turns ratio of the 16A3 transformer.

   $V_{IN} = $ _____

   $V_{OUT} = $ _____

   Turns ratio $= \dfrac{V_{IN}}{V_{OUT}}$

   $= $ _____ : _____

6. Find the turns ratio of a small power transformer supplied by your instructor.

   $V_{IN} = $ _____

   $V_{OUT} = $ _____

   Turns ratio $= \dfrac{V_{IN}}{V_{OUT}}$

   $= $ _____ : _____

7. What method can you use to determine whether or not the center tap of any of the transformers is actually at the center of the winding?

   _____

   _____

   _____

   _____

   _____

   _____

Summing Up

In your notebook, record any observations, problems, and conclusions for this activity.

**Activity 22-2:
Transformer Primary-
Secondary Voltage Phase**

Name _____

Date _____

Class _____ Score _____

Discussion

The way the primary and secondary windings of a transformer are wound around the core determines the polarity of the voltage produced at the output of the secondary. Voltages in series are added; therefore, if two transformer windings are connected in series, the resulting total output will be the sum of the two voltages. If one of the transformers is connected in reverse so the voltages are out of phase, the resulting output voltage will be the difference between the two voltages. This is because connecting the windings to the opposite polarity is the same as reversing the winding around the core.

Objective

In this activity, you will investigate the primary-secondary voltage phase of a transformer.

Materials and Equipment

2–16A3 or similar transformers
1–Audio generator
1–Breadboard
1–Oscilloscope

Procedures

1. Place the two transformers in a breadboard and identify the primary and secondary terminals, as shown in **Figure 22-2A.** The terminals can be marked on a strip of masking tape placed across the edge of the breadboard.

2. Connect the generator to the primaries, and adjust the frequency to 1 kHz. Measure and record the secondary voltages shown in Figure 22-2A.

   V_{1-3} = _____

   V_{4-5} = _____

3. Connect transformer terminals 3 and 4 together, as shown in **Figure 22-2B.** Measure and record voltage V_{1-5} in Figure 22-2B. If the voltage is zero, reverse the connections of terminals C and D. Mark the phase sense on transformer terminals A and C.

   V_{1-5} = _____

4. Connect transformer terminals 2 and 4 together, as shown in **Figure 22-2C.** Measure and record voltage V_{1-5}.

   V_{1-5} = _____

5. Explain the measurement result.

   _____

   _____

(Continued)

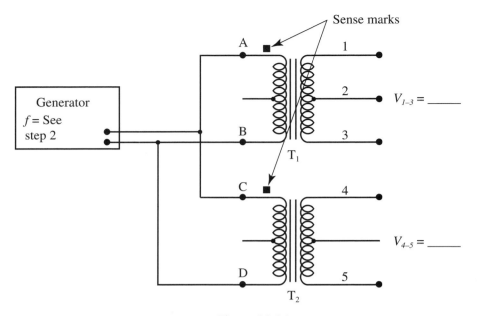

Figure 22-2A

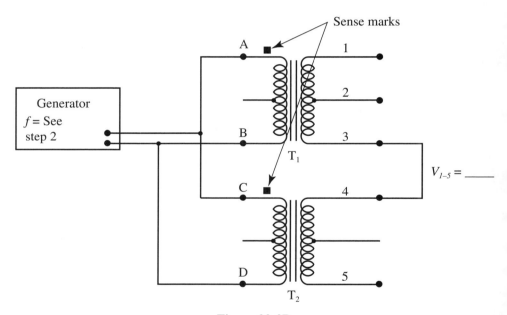

Figure 22-2B

6. Change the primary C and D terminal connections of transformer T_2, as shown in **Figure 22-2D.** This changes the phase sense mark and makes a change of 180° in the phase of the primary and secondary windings of transformer T_2. Measure and record voltage $V_{1-5.}$

   $V_{1-5} =$ _____

7. Explain the measurement result.

   _____

   _____

8. Change the transformer terminal connections by connecting 3 and 4, as shown in **Figure 22-2E.** Measure and record voltage $V_{1-5.}$

   $V_{1-5} =$ _____

(Continued)

Name _____

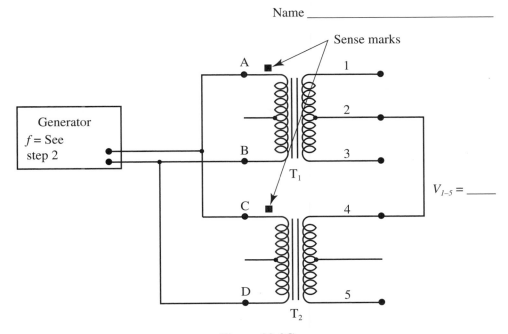

Figure 22-2C

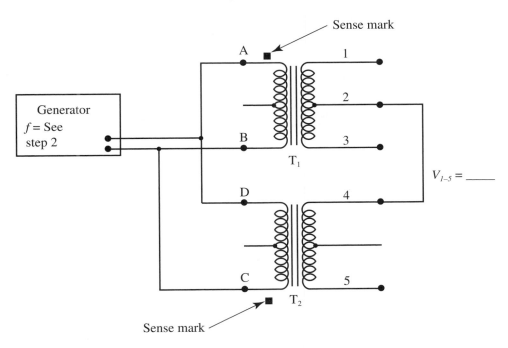

Figure 22-2D

9. Explain why the measurement is zero.

_____

_____

10. Change the primary A and B terminal connections of transformer T_1, as shown in **Figure 22-2F.** This changes the phase sense mark and makes a change of 180° in the phase of the primary and secondary windings of transformer T_1. Measure and record voltage V_{1-5}.

$V_{1-5} =$ _____

(Continued)

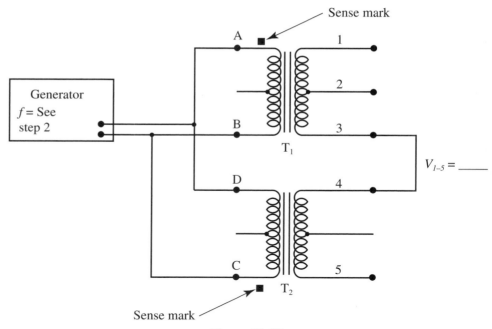

Figure 22-2E

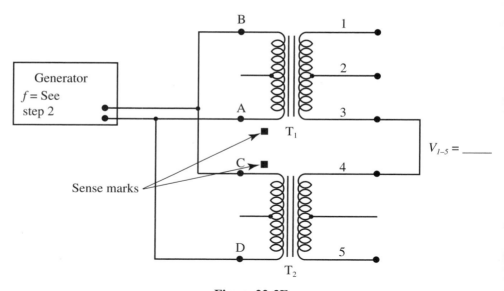

Figure 22-2F

11. Explain the measurement result.

_____

_____

12. Connect transformer terminals 3 and 5 together, as shown in **Figure 22-2G.**
    Measure and record voltage V_{1-4}.

    V_{1-4} = _____

13. Explain the measurement result.

_____

_____

(Continued)

Name _____

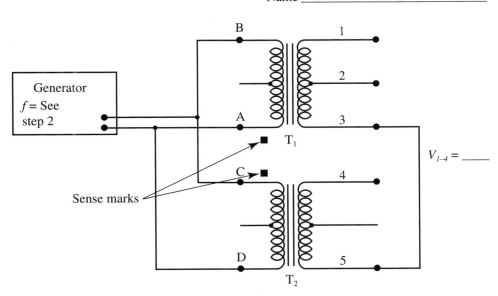

Figure 22-2G

Summing Up

In your notebook, record any observations, problems, and conclusions for this activity.

CAPACITANCE ▶ 23

Activity 23-1:
Capacitor Operation
with Ac and Dc

Name _____

Date _____

Class _____ Score _____

Discussion

A capacitor consists of two metal plates separated by an insulator. It stores an electrical charge. This charge occurs when electrons are redistributed from one side of the capacitor to the other. Capacitors are used to pass signals between amplifier stages and to block the flow of direct current through parts of an electronic circuit.

Objective

In this activity, you will investigate capacitor operation with alternating and direct currents.

Materials and Equipment

One each of the following capacitors:
   0.001 μF
   0.01 μF
   0.1 μF
   1 μF
1–1K Ω resistor
1–Audio generator
1–Breadboard
1–Dc power supply
1–Oscilloscope
1–VOM

Procedures

Part 1: Ac Operation

1. Build the circuit shown in **Figure 23-1A.**

2. Measure the voltage across resistor R_1 and record it in the table in **Figure 23-1B.**

3. Change C1 to the next value in the table and repeat step 2.

4. Continue this process with the remaining capacitor values.

Part 2: Dc Operation

1. Build the circuit shown in **Figure 23-1C.**

(Continued)

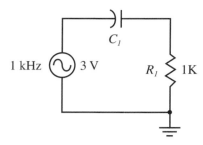

	C_I	V_R
	0.001 µF	
	0.01 µF	
	0.1 µF	
	1 µF	

Figure 23-1A   **Figure 23-1B**

2. Measure the voltage across resistor R_I and record it in the table in **Figure 23-1D.**

3. Change C_I to the next value in the table and repeat step 2.

4. Continue this process with the remaining capacitor values.

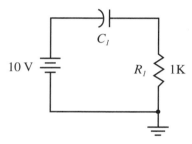

	C_I	V_R
	0.001 µF	
	0.01 µF	
	0.1 µF	
	1 µF	

Figure 23-1C   **Figure 23-1D**

Part 3: Questions

1. What are the differences between ac and dc operation of a capacitor?

_____

_____

2. How do you explain the results of the dc measurements?

_____

_____

Summing Up ─────────────────────────────

In your notebook, record any observations, problems, and conclusions for this activity.

Activity 23-2:
Charging and Discharging
Time Constant

Name _____

Date _____

Class _____ Score _____

Discussion

The time constant indicates the rate of charge or discharge of a capacitor and is determined by the product of the resistance and capacitance of the circuit. When a capacitor is charging, RC specifies the time it takes the capacitor to charge to 63.2% of the charging voltage. The higher the capacitance, the longer the time required to charge the capacitor. A small capacitance can produce time constants of milliseconds or picoseconds. Time constants are used in timing circuits such as waveform generation.

Objective

In this activity, you will investigate the charging and discharging time constant of a capacitor.

Materials and Equipment

One each of the following capacitors:
   47 µF
   100 µF
   1000 µF
   2200 µF
1–1K Ω resistor
1–Breadboard
1–Dc power supply
1–VOM
1–Watch with a second hand

Procedures

1. Construct the circuit in **Figure 23-2A.** Make sure switch S_1 is in the open position.

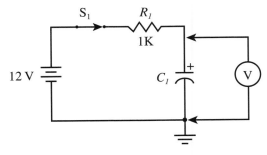

Figure 23-2A

2. Calculate 63.2% of the maximum voltage. This is the voltage to which the capacitor will charge during the first time constant.

   $V_{PER} =$ _____

(Continued)

Note: During the following steps, it may be helpful if another person keeps track of the time for you.

3. Using the 2200 μF capacitor and the 47K Ω resistor, close switch S_1, and time the voltage to the point calculated in step 2. Record the time in the table in **Figure 23-2B.**

	Time		
C_I	**47K**	**22K**	**10K**
2200 μF			
1000 μF			
100 μF			
47 μF			

Figure 23-2B

4. Return S_1 to the OFF position. Turn off the power and connect a shorting test lead across the capacitor to fully discharge it.

5. Change the resistor to 22K Ω and repeat steps 3 and 4, recording the results in the table.

6. Using steps 3 and 4, continue with the remaining capacitor-resistor combinations listed in the table.

7. How does the time constant of the larger capacitive values compare with the smaller values?

_____

_____

_____

Summing Up ━━━━━━━━━━━━━━━━━━━━━━━━━━━━

In your notebook, record any observations, problems, and conclusions for this activity.

RC CIRCUITS 24

Activity 24-1:
RC Circuit Operation

Name _____

Date _____

Class _____ Score _____

Discussion

Resistance changes in an RC circuit cause the impedance and the voltage-current phase angle to change. Because the voltage and current in a resistor are in phase, the voltage across a resistor is representative of its current. In this activity, the source voltage is compared to the voltage across the resistance, which is caused by the current flowing through the resistor. Therefore, the source voltage is being compared to the current in the circuit.

Objective

In this activity, you will investigate the effects of resistance in a capacitive circuit and the related phase angle.

Materials and Equipment

One each of the following resistors:
1500 Ω
4700 Ω
10K Ω
33K Ω
100K Ω
1–0.001 μF capacitor
1–Audio generator
1–Breadboard

Procedures

1. Connect the circuit shown in **Figure 24-1A** with the 1500 Ω resistor.

2. Set the generator for 3/4 maximum output voltage.

3. Set the generator for 20 kHz.

4. Connect channel 1 of the oscilloscope on the source side of the inductor and channel 2 at the top side of the resistor.

5. Using a method for good accuracy, adjust the oscilloscope controls to measure phase angle.

6. Measure the phase angle of the two voltages and record them in the table in **Figure 24-1B.**

(Continued)

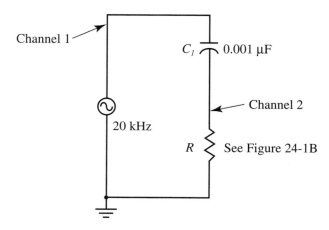

Figure 24-1A

R	Divisions out of Phase	Degrees out of Phase	Calculated Phase Angle
1500			
4700			
10K			
33K			
100K			

Figure 24-1B

7. Replace the resistor with the next resistor indicated in the table, and measure the phase angle again.

8. Repeat for the remaining resistors in the table.

9. Calculate the phase angle for each resistance in the table.

10. How do the measured and calculated phase angles compare?

_____

_____

11. On a separate sheet of paper, draw a vector diagram for each of the phase conditions in the table. (Optional)

Summing Up

In your notebook, record any observations, problems, and conclusions for this activity.

Activity 24-2:
RC Integration and
Differentiation

Name _____

Date _____

Class _____ Score _____

Discussion

Integrator and differentiator are circuits whose output is a mathematical form of calculus. The frequency causes changes in the output because of the time constant changes. These types of circuits are used in such applications as computers, robotics, lasers, and communications equipment. This activity explores integrator and differentiator circuits and shows how the output changes with frequency.

Objective

In this activity, you will investigate integration and differentiation in an RC circuit.

Materials and Equipment

1–0.001 µF capacitor
1–10K Ω resistor
1–33K Ω resistor
1–Audio generator
1–Breadboard

Procedures

Part 1: Integration

1. Connect the circuit as shown in **Figure 24-2A.**

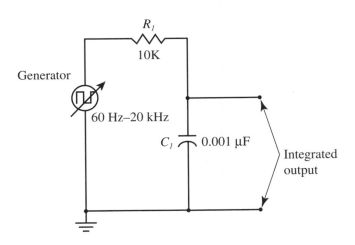

Figure 24-2A

2. Set the generator for 500 Hz.

3. Draw the waveform of the integrated output on Graph A in **Figure 24-2B.**

4. Change the generator frequency to the next frequency given in Graph B and draw the output waveform.

(Continued)

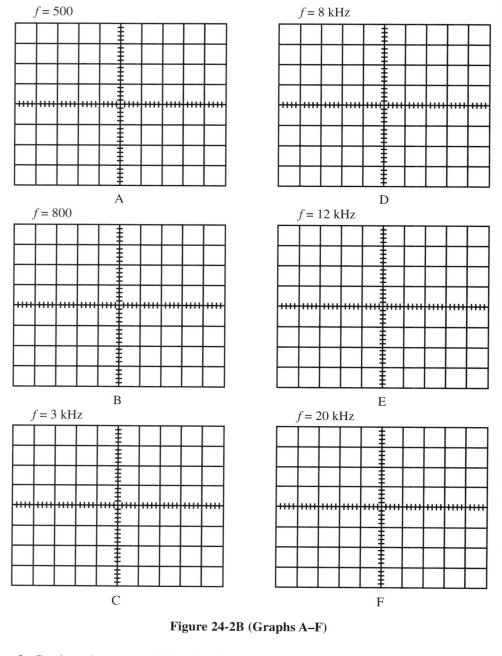

Figure 24-2B (Graphs A–F)

5. Continue the process of changing the generator frequency and drawing the waveform for all the graphs in Figure 24-2B.

6. At what frequency does the output become an exponential waveform? _____

7. How does the increase in frequency effect the charging and discharging of the capacitor?

_____

_____

Part 2: Differentiation

1. Change the circuit as shown in **Figure 24-2C.**

2. Adjust the generator for 20 kHz.

3. Draw the waveform of the integrated output on Graph A of **Figure 24-2D.**

(Continued)

Name _____

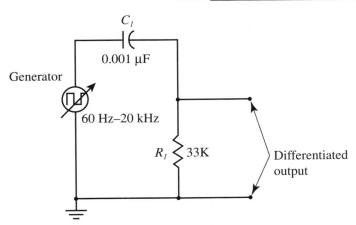

Figure 24-2C

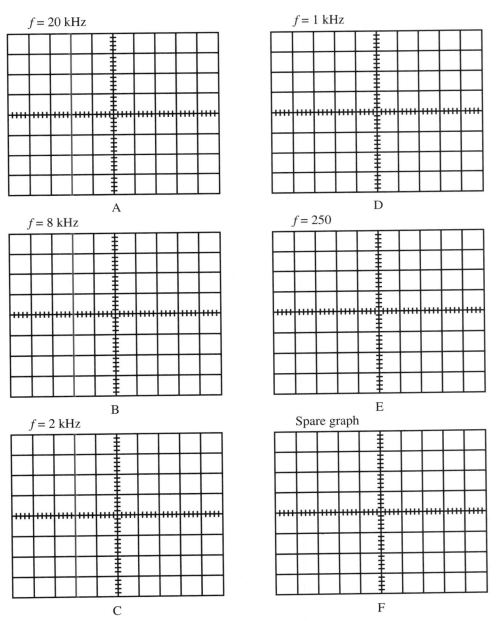

Figure 24-2D (Graphs A–F)

(Continued)

4. Change the generator frequency to the next frequency given in Graph B and draw the output waveform.

5. Continue the process of changing the generator frequency and drawing the waveform for all the graphs in Figure 24-2D.

6. How does the decrease in frequency effect the charging and discharging of the capacitor?

_____

_____

Summing Up ─────────────────────────────────

In your notebook, record any observations, problems, and conclusions for this activity.

RCL CIRCUITS ➤ 25

Activity 25-1:
Series Resonant Circuits

Name _____

Date _____

Class _____ Score _____

Discussion

RF circuits, which are used for tuning a signal at a specific frequency, are a common application of RCL circuits. Tuning is used in radio/television transmitters and receivers as well as antenna circuits. At resonance, the net reactance is zero; that is, the inductive reactance is equal to the capacitive reactance at resonance. This means the impedance is equal to the resistance of the circuit, causing line current of the circuit to be maximum. In this activity, the maximum line current through a series resistance provides a voltage drop to indicate resonance.

Objective

In this activity, you will investigate the operation of series resonant circuits.

Materials and Equipment

1–47 pF capacitor
1–0.5 mH inductor
1–1K Ω resistor
1–AM radio receiver
1–Breadboard
1–Frequency counter (optional)
1–Oscilloscope
1–RF generator

Procedures

Part 1

1. Connect the series circuit as shown in **Figure 25-1.**

2. Calculate the resonant frequency.

   Calculated f_r = _____

3. Adjust the generator for the calculated frequency.

4. Using an oscilloscope, observe the voltage across resistor R_1 and adjust the RF generator to circuit resonance.

5. Measure the resonant frequency.

   Measured f_r = _____

(Continued)

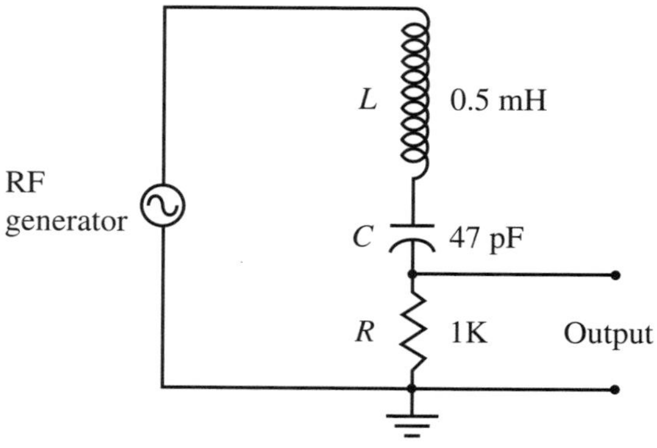

Figure 25-1

6. Using only the resonant frequency band on the generator, draw a frequency-to-voltage response curve for the circuit.

7. What are the indications that the circuit is resonant?

_____

_____

_____

Part 2

1. Place an AM radio receiver 1′ to 2′ from the circuit and turn on the radio.

2. Tune the radio to the resonant frequency.

3. What are the indications that the circuit is resonant?

_____

_____

_____

Summing Up

In your notebook, record any observations, problems, and conclusions for this activity.

Activity 25-2:	Name _____
Parallel Resonant Circuits	Date _____
	Class _____ Score _____

Discussion

As with series RCL circuits, parallel circuits are used extensively in RF tuning circuits. The bandwidth of the circuit can be controlled by series and parallel shunt resistances which determine the Q of the circuit. In this activity, both the damped and undamped parallel RCL circuit are used.

Objective

In this activity, you will investigate the operation of parallel resonant circuits.

Materials and Equipment

1–47 pF capacitor
1–0.5 mH inductor
1–1K Ω resistor
1–6.8K Ω resistor
1–Breadboard
1–Frequency counter (optional)
1–RF generator
1–Oscilloscope

Procedures

Part 1: Undamped Parallel RCL Circuit

1. Connect the parallel circuit as shown in **Figure 25-2A.**

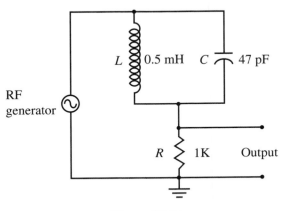

Figure 25-2A

2. Calculate the resonant frequency.

   Calculated $f_r =$ _____

3. Adjust the generator for the calculated frequency.

4. Using an oscilloscope, observe the voltage across resistor R_1 and adjust the RF generator to resonance.

5. Measure the resonant frequency.

   Measured $f_r =$ _____

(Continued)

6. Using only the resonant frequency band on the generator, draw a frequency-to-voltage response curve for the circuit.

7. What are the indications that the circuit is resonant?

_____

_____

Part 2: Damped Parallel RCL Circuit

1. Connect the parallel circuit as shown in **Figure 25-2B,** which adds the parallel 6.8K Ω resistor R_2 to the circuit.

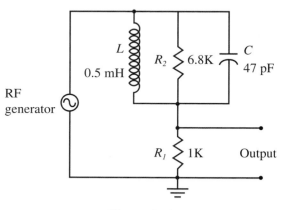

Figure 25-2B

2. Calculate the resonant frequency.

Calculated $f_r =$ _____

3. Adjust the generator for the calculated frequency.

4. Using an oscilloscope, observe the voltage across resistor R_1, and adjust the RF generator to resonance.

5. Measure the resonance frequency.

Measured $f_r =$ _____

6. Using only the resonant frequency band on the generator, draw a frequency-to-voltage response curve for the circuit.

7. Change resistor R_2 to a higher or lower value, and note the change in the circuit response. (Optional)

8. What are the indications that the circuit is resonant?

_____

9. In what way is the damped parallel resonant circuit different from the undamped parallel resonant circuit?

_____

10. In what way is the parallel resonant circuit different from the series resonant circuit?

_____

Summing Up ─────────────────────────────────

In your notebook, record any observations, problems, and conclusions for this activity.

FILTER CIRCUITS 26

Activity 26-1:
Low-pass Filter Circuits

Name _____

Date _____

Class _____ Score _____

Discussion

Simple RC and RL circuits can be used in circuit design to adjust the frequency response. When the response increases in a high-pass filter or decreases in a low-pass filter, the frequency cutoff point f_{CO} is visible at 3 dB down from maximum. The f_{CO} is calculated by $f_{CO} = 1 / 2\pi RC$. A change in resistance or capacitance will change the time constant and the frequency cutoff point.

Objectives

In this activity, you will:
- Demonstrate the operation of a low-pass filter circuit.
- Measure the frequency response of an RC low-pass filter.

Materials and Equipment

One each of the following Mylar™ capacitors:
   0.033 µF
   0.01 µF
   0.22 µF
   0.47 µF
1–1.2K Ω resistor
1–Audio generator
1–Breadboard
1–Oscilloscope

Procedures

1. Using a 0.033 µF Mylar capacitor, connect the circuit as shown in **Figure 26-1A.**

2. Adjust the potentiometer shaft for the approximate midpoint of its rotation. The potentiometer is necessary to maintain a constant input voltage to the filter circuit.

3. Adjust the generator for 200 Hz and the generator output amplitude for a convenient input voltage V_{IN}, such as 2 $V_{P\text{-}P}$. This input voltage amplitude must be maintained throughout the remaining frequencies.

   $V_{IN} = $ _____

4. Measure the peak-to-peak output voltage V_O, and record it in Table A of **Figure 26-1B.**

(Continued)

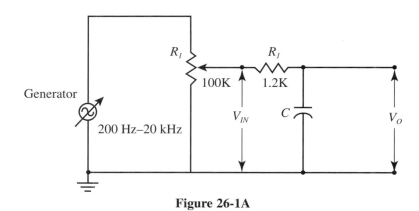

Figure 26-1A

5. Adjust the generator for the next frequency listed in Table A.

6. Adjust potentiometer R_1 for the constant input voltage measured in step 2.

7. Measure the peak-to-peak output voltage V_O and record it in Table A.

$C = 0.01\ \mu F$

Frequency	V_O	dB Response
200 Hz		
500 Hz		
1 kHz		
2 kHz		
3 kHz		
4 kHz		
5 kHz		
7.5 kHz		
10 kHz		
15 kHz		
20 kHz		

Table A

$C = 0.033\ \mu F$

Frequency	V_O	dB Response
200 Hz		
500 Hz		
1 kHz		
2 kHz		
3 kHz		
4 kHz		
5 kHz		
7.5 kHz		
10 kHz		
15 kHz		
20 kHz		

Table B

$C = 0.22\ \mu F$

Frequency	V_O	dB Response
200 Hz		
500 Hz		
1 kHz		
2 kHz		
3 kHz		
4 kHz		
5 kHz		
7.5 kHz		
10 kHz		
15 kHz		
20 kHz		

Table C

$C = 0.47\ \mu F$

Frequency	V_O	dB Response
200 Hz		
500 Hz		
1 kHz		
2 kHz		
3 kHz		
4 kHz		
5 kHz		
7.5 kHz		
10 kHz		
15 kHz		
20 kHz		

Table D

Figure 26-1B (Tables A–D)

(Continued)

Name _____

8. Using Steps 4 through 6, continue this process for the remaining frequencies in Table A.

9. Using the equation $dB = 20 \log \dfrac{V_2}{V_1}$, calculate and record the dB response for each frequency in Table A.

10. Using semilog graph paper, plot the response curve in decibels.

11. Using the preceding process, change the capacitor to 0.01 µF and complete Tables B, C, and D. For comparison, plot the response curve for each on the same graph paper used in step 10.

12. Calculate the frequency cutoff for each of the following capacitances:

   $C = 0.033 \ \mu F$ $f_{CO} =$ _____

   $C = 0.01 \ \mu F$ $f_{CO} =$ _____

   $C = 0.22 \ \mu F$ $f_{CO} =$ _____

   $C = 0.47 \ \mu F$ $f_{CO} =$ _____

13. How did the different capacitor values affect the frequency cutoff point?

   _____

   _____

   _____

14. If the input signal voltage is not held constant, how will this affect the results of the response curve?

   _____

   _____

   _____

Summing Up ────────────────────────────────

In your notebook, record any observations, problems, and conclusions for this activity.

Activity 26-2:	Name _____
High-pass Filter Circuits	Date _____
	Class _____ Score _____

Discussion

This activity demonstrates the high-pass filter and compares it to the low-pass filter of the preceding activity. On a response curve, the frequency cutoff point (f_{CO}) is shown at 3 dB down from maximum on a response curve. As before, $f_{CO} = 1 / 2\pi\, RC$.

Objectives

In this activity, you will:
- Demonstrate the operation of a high-pass filter circuit.
- Measure the frequency response of an RC high-pass filter.

Materials and Equipment

One each of the following Mylar™ capacitors:
  0.033 µF
  0.01 µF
1–3.3K Ω resistor
1–Audio generator
1–Breadboard
1–Oscilloscope

Procedures

1. Connect the circuit as shown in **Figure 26-2A** using a 0.033 µF Mylar capacitor.

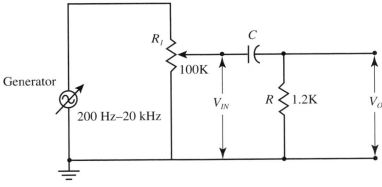

Figure 26-2A

2. Adjust the potentiometer shaft for the approximate midpoint of its rotation. The potentiometer is necessary to maintain a constant input voltage to the filter circuit.

3. Adjust the generator for 200 kHz and the generator output amplitude for a convenient input voltage (V_{IN}), such as 2 V_{P-P}. This input voltage amplitude must be maintained throughout the remaining frequencies.

   $V_{IN} = $ _____

(Continued)

4. Measure the peak-to-peak output voltage (V_O) and record it in Table A of **Figure 26-2B.**

5. Adjust the generator for the next frequency in Table A.

6. Adjust potentiometer R_1 for the constant input voltage measured in step 2.

7. Measure the peak-to-peak output voltage (V_O) and record it in Table A.

8. Repeat Steps 4 through 6 for the remaining frequencies in Table A.

9. Using the equation $dB = 20 \log \dfrac{V_2}{V_1}$, calculate and record the dB response for each frequency in Table A.

10. Using semilog graph paper, plot the response curve in decibels.

$C = 0.01\ \mu F$

Frequency	V_O	dB Response
200 Hz		
500 Hz		
1 kHz		
2 kHz		
3 kHz		
4 kHz		
5 kHz		
7.5 kHz		
10 kHz		
15 kHz		
20 kHz		

Table A

$C = 0.33\ \mu F$

Frequency	V_O	dB Response
200 Hz		
500 Hz		
1 kHz		
2 kHz		
3 kHz		
4 kHz		
5 kHz		
7.5 kHz		
10 kHz		
15 kHz		
20 kHz		

Table B

Figure 26-2B (Tables A and B)

11. Following the preceding process, change the capacitor to 0.01 µF and complete Table B. For comparison, plot the response curve for each on the same graph paper used in step 10.

12. Calculate the frequency cutoff for each of the following capacitances:

    $C = 0.033\ \mu F$ $f_{CO} = \underline{\hspace{2cm}}$

    $C = 0.01\ \mu F$ $f_{CO} = \underline{\hspace{2cm}}$

13. How did the different capacitor values affect the frequency cutoff point?

_____

_____

_____

14. How does the high-pass filter compare with the low-pass filter of Activity 26-1?

_____

_____

_____

Summing Up ————————————————————————

In your notebook, record any observations, problems, and conclusions for this activity.

COMPLEX CIRCUIT ANALYSIS 27

Activity 27-1:
RCL Circuit Analysis

Name _____

Date _____

Class _____ Score _____

Discussion

Any ac circuit can be analyzed using complex numbers, especially combination series-parallel circuits that involve both resistances and reactances. In this activity, you will calculate the currents, voltages, and impedances of the circuit and measure the voltages for comparison.

Objective

In this activity, you will analyze a complex RCL circuit.

Materials and Equipment

One each of the following Mylar™ capacitors:
   0.1 μF
   0.22 μF
One each of the following resistors:
   100 Ω
   150 Ω
   470 Ω
One each of the following inductors:
   1 mH
   2.5 mH
1–AF generator
1–Breadboard
1–Oscilloscope

Procedures

1. Connect the circuit as shown in **Figure 27-1A.**

2. Adjust the generator for 2 V_{P-P} at 20 kHz.

   Note: The circuit has a low impedance and, depending on the generator output capabilities, may load the generator output excessively. A voltage other than 2 V can be used. Simply record it and continue.

   $V_S =$ _____

3. Using complex numbers, calculate the impedances and currents in the table in **Figure 27-1B.** Remember to consider the source measured in step 2.

(Continued)

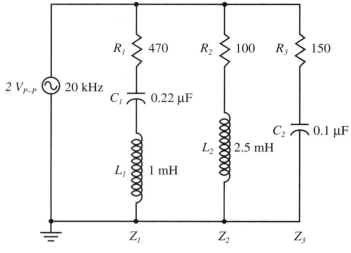

Figure 27-1A

	Calculated		
Z_1		I_1	
Z_2		I_2	
Z_3		I_3	
Z_T		I_T	

Figure 27-1B

4. Calculate and record the voltages in the table in **Figure 27-1C.**

5. Using an oscilloscope, measure and record the voltages in the table.

Quantity	Calculated	Measured
V_{L_1}		
V_{L_2}		
V_{C_2}		
V_{R_1}		
V_{R_2}		
V_{R_3}		
$V_{L_1-C_1}$		
V_S		

Figure 27-1C

6. With the ground connected to the other side of the circuit, change the circuit as shown in **Figure 27-1D.** This will allow measurement of the voltages across the resistors.

7. Using the oscilloscope, measure the voltages across the resistors in the table in Figure 27-1C.

(Continued)

Name _____

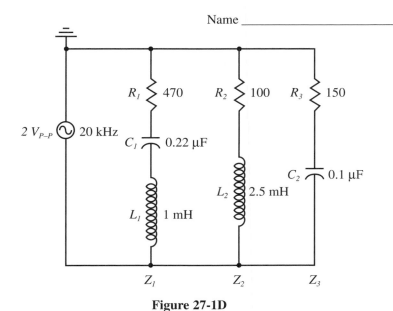

Figure 27-1D

8. Using the measured voltages, calculate the source voltage across each circuit branch.

$Z_1 : V_S =$ _____

$Z_2 : V_S =$ _____

$Z_3 : V_S =$ _____

9. How do the calculated voltages compare with the measured values?

_____

_____

_____

10. How do you explain the differences?

_____

_____

_____

Summing Up

In your notebook, record any observations, problems, and conclusions for this activity.

60 HERTZ AND THREE-PHASE AC POWER

Activity 28-1:
Ladder Diagram Project

Name _____

Date _____

Class _____ Score _____

Discussion

A ladder diagram helps you understand how any 60 Hz power system is wired. This information is necessary to determine if the power supply system is sufficient. By drawing a ladder diagram of the wiring, the power and current to be used by the electrical equipment can be calculated.

Objective

In this activity, you will make a ladder diagram of a wired room.

Materials and Equipment

1–Graph paper
1–Pencil
1–Ruler

Procedures

1. Using proper electrical symbols, make a ladder diagram of the electrical system assigned by your instructor.

2. Identify the fuse or circuit breaker for each part of the diagram.

3. Identify the watts of power for each load on the diagram.

4. Calculate the current for each part of the diagram.

5. Calculate the total maximum power required by each part of the diagram.

6. How do the calculated powers and currents compare with the fused circuit values?

_____

_____

_____

7. Is the fuse or circuit breaker system adequate ?

_____

_____

(Continued)

Summing Up

In your notebook, record any observations, problems, and conclusions for this activity.

ELECTRIC MOTORS 29

Activity 29-1:
Motor Information

Name _____

Date _____

Class _____ Score _____

Discussion

A vast number of electric motors are in use today. Vital information about parts and operation of motors must be understood.

Objective

In this activity, you will review important information about motors.

Questions

1. What is a locked rotor?

   _____

   _____

2. Identify the parts of the motor shown in **Figure 29-1.**

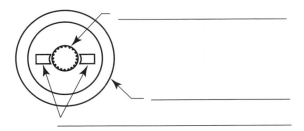

Figure 29-1

3. List four possible problems with motors.

   _____

   _____

   _____

   _____

(Continued)

4. List the five types of dc motors.

　　　_____

　　　_____

　　　_____

　　　_____

　　　_____

5. A shaded-pole motor operates on what kind of current?

　　　_____

6. Maintenance of small motors consists of:

　_____

　_____

　_____

7. What are four reasons a motor would run hot?

　_____

　_____

　_____

　_____

8. A motor shaft is hard to turn by hand. What is the cause?

　_____

9. A noisy motor is usually caused by:

　_____

10. Important items found on a motor name plate are:

　　　_____

　　　_____

　　　_____

　　　_____

　　　_____

　　　_____

Summing Up

In your notebook, record any observations, problems, and conclusions for this activity.

MOTOR CONTROLS ◄ 30

Activity 30-1:
Logic Functions

Name _____

Date _____

Class _____ Score _____

Discussion

In the **AND** function, the lamp is on only when both switch 1 AND switch 2 are in the ON position. In the **OR** function, the lamp is on when either switch 1 OR switch 2 is in the ON position.

In the **NOT** function, an output is present when the control signal is off. Combining the NOT function with the AND function produces a NAND logic function. In the **NAND** function, the lamp is de-energized when both switches are open. Likewise, the **NOR** logic is a combination of NOT logic and OR logic. Opening either switch 1 OR switch 2 will cause NO output.

Objective

In this activity, you will demonstrate the logic functions AND, OR, NAND, and NOR.

Materials and Equipment

1–120 V ac/6.3 V ac transformer
1–#47 pilot lamp with socket
1–Breadboard
2–SPST N.C. push-button switches
2–SPST N.O. push-button switches

Procedures

Warning! 120 V ac power is dangerous.
Part 1

1. Connect the wires on the push-button switches so they can be connected to the breadboard.

2. Connect the circuit as shown in **Figure 30-1A.**

3. Connect the transformer to 120 V ac.

4. With neither push-button switch pressed (both off), write the results in the table in Figure 30-1A.

5. Press switch PB_1 and write the results in the table. Release switch PB_1.

6. Continue the switch combinations given in the table.

(Continued)

7. When does the lamp give output?

_____

8. What logic function is being performed?

_____

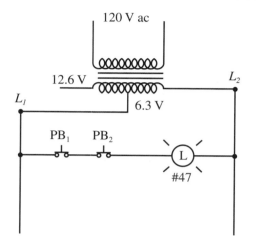

Figure 30-1A

Signal		Output
PB$_2$	PB$_1$	L
Off	Off	
Off	On	
On	Off	
On	On	

Part 2

1. Connect the circuit as shown in **Figure 30-1B.**

2. Connect the transformer to 120 V ac.

3. With neither push-button switch pressed (both off), write the results in the table in Figure 30-1B.

4. Press switch PB$_1$ and write the results in the table. Release switch PB$_1$.

5. Continue the switch combinations given in the table.

6. When does the lamp give output?

_____

7. What logic function is being performed?

_____

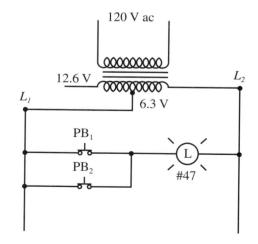

Figure 30-1B

Signal		Output
PB$_2$	PB$_1$	L
Off	Off	
Off	On	
On	Off	
On	On	

(Continued)

Name _____

Part 3

1. Connect the circuit as shown in **Figure 30-1C.**

2. Connect the transformer to 120 V ac.

3. With neither push-button switch pressed (both off), write the results in the table in Figure 30-1C.

4. Press switch PB_1 and write the results in the table. Release switch PB_1.

5. When does the lamp give output?

_____

6. What logic function is being performed?

_____

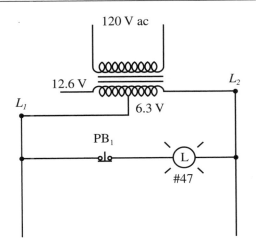

Signal	Output
PB_1	L
Off	
On	

Figure 30-1C

Part 4

1. Connect the circuit as shown in **Figure 30-1D.**

2. Connect the transformer to 120 V ac.

3. With neither push-button switch pressed (both off), write the results in the table in Figure 30-1D.

4. Press switch PB_1 and write the results in the table. Release switch PB_1.

5. Continue the switch combinations given in the table.

6. When does the lamp give output?

_____

7. What logic function is being performed?

_____

Part 5

1. Connect the circuit as shown in **Figure 30-1E.**

2. Connect the transformer to 120 V ac.

3. With neither push-button switch pressed (both off), write the results in the table in Figure 30-1E.

(Continued)

4. Press switch PB$_1$ and write the results in the table. Release switch PB$_1$.

5. Continue the switch combinations given in the table.

6. When does the lamp give output?

_____

7. What logic function is being performed?

_____

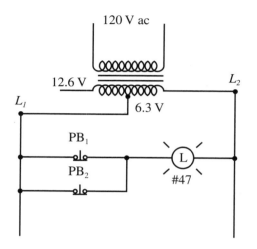

Signal		Output
PB$_2$	PB$_1$	L
Off	Off	
Off	On	
On	Off	
On	On	

Figure 30-1D

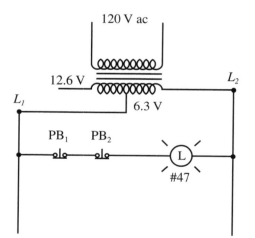

Signal		Output
PB$_2$	PB$_1$	L
Off	Off	
Off	On	
On	Off	
On	On	

Figure 30-1E

Summing Up

In your notebook, record any observations, problems, and conclusions for this activity.

<table>
<tr><td rowspan="3">**Activity 30-2:**
Memory Logic Function</td><td>Name _____</td></tr>
<tr><td>Date _____</td></tr>
<tr><td>Class _____ Score _____</td></tr>
</table>

Discussion

Memory logic is used when a circuit must remember what the input signal was after the signal is removed. When a circuit is on, it remains on until it is turned off by another signal, even though the ON signal was removed. Memory logic is accomplished by mechanical relay contacts or a solid-state relay, which keeps the circuit in the ON state until an OFF signal is received.

Objective

In this activity, you will demonstrate the memory logic functions.

Materials and Equipment

1–120 V ac pilot lamp with socket
1–Breadboard
1–Relay, Radio Shack #275-217 with socket
1–SPST N.O. push-button switches, Radio Shack #275-1547
1–SPST N.C. push-button switches, Radio Shack #275-1548

Procedures

Caution! 120 V ac power is dangerous.

1. Connect wires on the push-button switches so they can be connected to the breadboard. Mount the switches in a metal box to eliminate shock hazard.

2. Connect the circuit as shown in **Figure 30-2.**

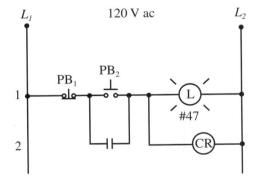

Figure 30-2

3. Press PB$_2$; then release the switch. What happened?

_____

_____

(Continued)

4. Press push-button switch PB$_1$; then release the switch. What happened?

_____

_____

5. How does the relay hold the circuit in the ON state?

_____

_____

Summing Up

In your notebook, record any observations, problems, and conclusions for this activity.

SEMICONDUCTOR FUNDAMENTALS

Activity 31-1:
Diode Voltage and Current

Name _____

Date _____

Class _____ Score _____

Discussion

Diodes are nonlinear devices. A voltage-current graph of a diode is not a straight (linear) line. In other words, the current is not directly proportional to its voltage. When the voltage across a silicon diode is below 0.7 V, only a small amount of current flows. Once the voltage goes above 0.7 V, the current increases rapidly. This activity investigates the voltage-current relationship of a silicon diode.

Objective

In this activity, you will investigate the operation of a diode with changing source voltage.

Materials and Equipment

1–1N4001 diode
1–470 Ω resistor
1–Breadboard
1–Variable dc power supply
1–VOM

Procedures

1. Connect the circuit as shown in **Figure 31-1A.**

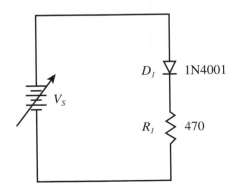

Figure 31-1A

(Continued)

2. Turn the power on, and adjust the voltage source for 4 V.

3. Measure the voltage drop across diode D_1 and resistor R_1. Record these voltages in the table in **Figure 31-1B.**

V_S	V_R	V_D	Calculated I
4			
6			
8			
10			
12			

Figure 31-1B

4. Adjust the voltage source for the next voltage in the table, and measure the voltages again. Record the voltages in the table.

5. Repeat step 4 for the remaining voltages in the table.

6. Using the voltage drop across resistor R_1, calculate the circuit current for each of the voltages. Record each current in the table.

7. What are the differences in the voltage drop across diode D_1 and resistor R_1?

_____

_____

_____

_____

8. What is the relationship between the diode voltage and the circuit current?

_____

_____

_____

Summing Up

In your notebook, record any observations, problems, and conclusions for this activity.

Activity 31-2:
Forward and Reverse Bias

Name _____

Date _____

Class _____ Score _____

Discussion

When a voltage is connected with the positive on the anode and negative on the cathode, current will flow through the diode. The P-N junction is forward biased. When the polarities are reversed, only a small leakage current flows. When a diode is reverse biased, no current flows in the reverse direction.

Objective

In this activity, you will investigate the operation of forward and reverse biased diodes.

Materials and Equipment

2–1N4001 diode
2–470 Ω resistor
1–Breadboard
1–Dc power supply
1–VOM

Procedures

1. Connect the circuit as shown in **Figure 31-2.**

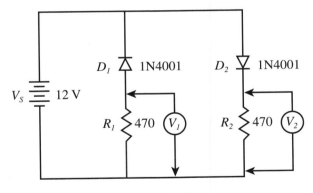

Figure 31-2

2. Adjust the voltage source for 12 V and turn on the power.

3. Measure voltage drops V_1 and V_2.

   $V_1 =$ _____

   $V_2 =$ _____

4. Calculate the current for each diode.

   $I_1 =$ _____

   $I_2 =$ _____

(Continued)

5. What are the differences between the two diode currents?

_____

_____

_____

6. Why are the two currents different?

_____

_____

_____

Summing Up

In your notebook, record any observations, problems, and conclusions for this activity.

Activity 31-3: Diode Effects on Alternating Current

Name _____

Date _____

Class _____ Score _____

Discussion

Alternating current changes polarity on each alternation of the cycle. When alternating current is applied to a diode, the diode is automatically forward and reverse biased on each alternation. The action of a diode on alternating current is called rectification, or the changing of ac to dc.

Objective

In this activity, you will investigate the effects of a diode on alternating current.

Materials and Equipment

1–1N4001 diode
1–1K Ω resistor
1–Audio generator
1–Breadboard
1–Oscilloscope

Procedures

1. Connect the circuit as shown in **Figure 31-3A.**

2. Adjust the frequency for 500 Hz at a readable amplitude level on the oscilloscope.

3. Observe the oscilloscope waveform.

4. Set the oscilloscope vertical input switch to dc.

5. Draw the waveform that appears on the graticule. See **Figure 31-3B.**

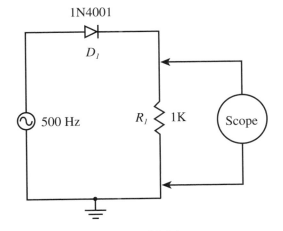

Figure 31-3A

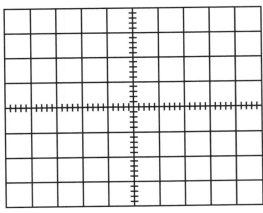

Figure 31-3B

(Continued)

6. How does the waveform react to the change of the input switch setting? Why does this action take place?

_____

_____

_____

_____

_____

7. Reverse the diode, as shown in **Figure 31-3C.**

8. Draw the waveform that appears on the graticule. See **Figure 31-3D.**

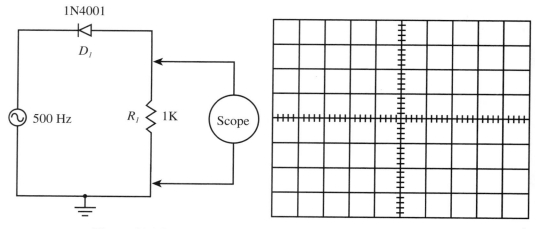

Figure 31-3C   **Figure 31-3D**

9. How does the waveform react to the change in position of the diode? Why does this action take place?

_____

_____

_____

_____

_____

Summing Up

In your notebook, record any observations, problems, and conclusions for this activity.

Activity 31-4:
Zener Diode Effects

Name _____

Date _____

Class _____ Score_____

Discussion

A zener diode operates much differently than an ordinary junction diode. When a junction diode reaches the breakdown voltage, an avalanche current flows in the reverse direction and the diode is damaged. On the other hand, when a zener diode reaches the breakdown voltage, it continues to operate perfectly within the circuit. The zener breakdown voltage gives the diode its ability to regulate the voltage. Zener diodes are manufactured with various voltages.

Objective

In this activity, you will investigate the operation of a zener diode.

Materials and Equipment

1–4700 Ω resistor
1–1NXXXX zener diode (see instructor)
1–Breadboard
1–Dc power supply
1–VOM

Procedures

1. Connect the circuit as shown in **Figure 31-4A.**

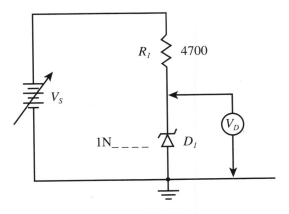

Figure 31-4A

2. Adjust the power supply for 4 V.

3. Measure the voltage across the resistor and zener diode. Record the measurements in the table in **Figure 31-4B.**

4. Adjust the power supply for each of the other voltages in the table. Measure the voltages and record the measurements.

5. Calculate the circuit current for each of the voltages in the table.

(Continued)

V_S	V_Z	I_T
4		
6		
8		
10		
12		
14		
18		

Figure 31-4B

6. How does voltage across the zener diode change compared to the voltage source and circuit current?

_____

_____

_____

_____

7. What is the zener voltage of this diode?

_____

Summing Up

In your notebook, record any observations, problems, and conclusions for this activity.

Activity 31-5: Testing Diodes

Name _____

Date _____

Class _____ Score _____

Discussion

A simple ohmmeter test is all that is needed to determine the quality of a diode. Diodes are seldom found to be open. They become shorted, causing blown fuses and other symptoms in the equipment. The ohmmeter and curve tracer methods are used in this activity to determine if a diode is good or defective.

Objective

In this activity, you will test diodes for good, open, or shorted conditions.

Materials and Equipment

1–6 V to 15 V filament transformer
1–2.2K Ω resistor
Assorted diodes
1–Breadboard
1–Oscilloscope with X-Y function key
1–VOM

Procedures

1. Set the ohmmeter (analog) on R×100 or the diode test position (digital).

2. As shown in **Figure 31-5A,** connect the meter to diodes 1 through 5. Test for good, open, or shorted condition.

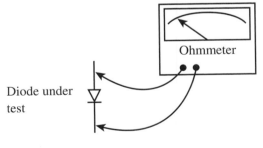

Diode under test

Figure 31-5A

3. Record the results in the ohmmeter column of the table in **Figure 31-5B.**

4. Set up the curve tracer as shown in **Figure 31-5C.**

5. Connect each of the diodes listed in the table. Record the results in the curve tracer column.

(Continued)

Diode	Ohmmeter	Curve Tracer
1		
2		
3		
4		
5		

Figure 31-5B

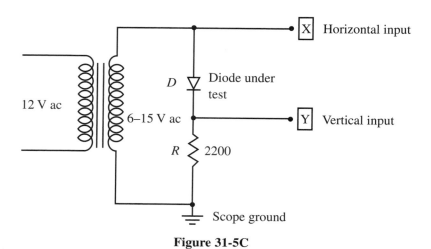

Figure 31-5C

6. How do the ohmmeter results compare with the curve tracer results?

_____

_____

_____

_____

_____

Summing Up

In your notebook, record any observations, problems, and conclusions for this activity.

DC POWER SUPPLIES 32

Activity 32-1:
Half-wave Power Supply

Name _____

Date _____

Class _____ Score _____

Discussion

A diode used in a power supply is called a rectifier because it changes ac to dc. When only one diode is used, it rectifies half of the ac cycle making it a half-wave power supply. A half-wave power supply is used for low-current applications such as battery chargers and ac-dc adaptors found in consumer electronic equipment. When a rechargeable battery is charged, a half-wave power supply charges the battery to the peak voltage of the secondary minus the forward voltage across the diode. The equation is $V_O = V_{PEAK} - V_F$.

Objective

In this activity, you will investigate the operation of a half-wave dc power supply.

Materials and Equipment

1–1N4001 diode
1–100 µF electrolytic capacitor
1–12.6 V filament transformer (center-tapped)
1–220 Ω, 2 W resistor
1–Breadboard
1–Oscilloscope
1–VOM

Procedures

1. Connect the circuit as shown in **Figure 32-1A.**

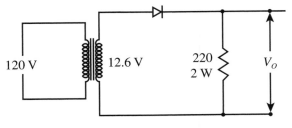

120 V 12.6 V 220 2 W V_O

Figure 32-1A

2. Connect the power transformer to ac power.

(Continued)

3. Turn on the power.

4. Using an oscilloscope, draw the waveform of the output across load resistor R_L in the graph in **Figure 32-1B.**

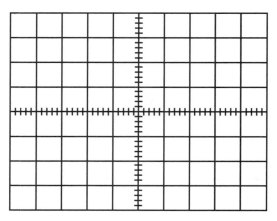

Figure 32-1B

5. What does the waveform measurement indicate about the circuit operation?

_____

_____

_____

6. Add the filter capacitor to the circuit. See **Figure 32-1C.**

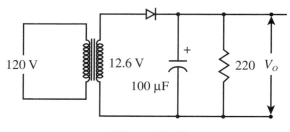

120 V   12.6 V   100 µF   220 V_O

Figure 32-1C

7. With a colored pen or pencil, draw the waveform of the output across the load resistor R_L on the graph above.

8. What difference do you see in the waveform? Why?

_____

_____

_____

_____

_____

Summing Up

In your notebook, record any observations, problems, and conclusions for this activity.

Activity 32-2:
Full-wave Center-tapped
Power Supply

Name _____

Date _____

Class _____ Score _____

Discussion

The advantage of a full-wave center-tapped power supply is it rectifies both alternations of the ac cycle. This provides more dc and reduces the amount of filtering needed. The disadvantage is the cost of a center-tapped transformer. Since only half the secondary is used at a time, the transformer voltage must be twice that needed to provide the dc output voltage. In addition, the diode PIV must be at least the peak value of the full secondary voltage.

Objective

In this activity, you will investigate the operation of a full-wave dc power supply.

Materials and Equipment

2–1N4001 diodes
1–100 μF electrolytic capacitor
1–12.6 V filament transformer (center-tapped)
1–220 Ω, 2 W resistor
1–Breadboard
1–Oscilloscope
1–VOM

Procedures

1. Connect the circuit as shown in **Figure 32-2A.**

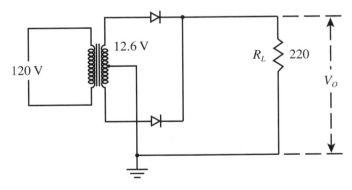

Figure 32-2A

2. Connect the power transformer to ac power.

3. Turn on the power.

4. Using an oscilloscope, draw the waveform of the output across the load resistor R_L in the graph in **Figure 32-2B.**

(Continued)

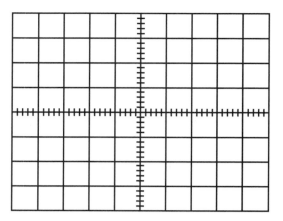

Figure 32-2B

5. What does the measured waveform indicate about the circuit operation?

_____

_____

6. Add the filter capacitor to the circuit. See **Figure 32-2C.**

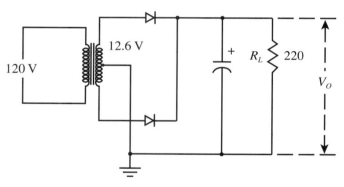

Figure 32-2C

7. Using a colored pen or pencil, draw the waveform of the output across load resistor R_L on the graph above.

8. What difference do you see in the waveform? Why?

_____

_____

_____

_____

_____

Summing Up

In your notebook, record any observations, problems, and conclusions for this activity.

<table>
<tr><td>Name</td></tr>
</table>

Activity 32-3:
Full-wave Bridge
Power Supply

Name _____

Date _____

Class _____ Score _____

Discussion

Because it uses four diodes in a bridge network, a bridge power supply eliminates the need for a center-tapped transformer. With slight changes, the bridge power supply can become dual polarity, supplying both a positive and negative dc power source. The peak inverse voltage (PIV) of the diodes must be greater than the peak secondary voltage.

Objective

In this activity, you will investigate the operation of a full-wave dc power supply.

Materials and Equipment

4–1N4001 diodes
1–12.6 V filament transformer (center-tapped)
1–220 Ω, 2 W resistor
1–Breadboard
1–Oscilloscope
1–VOM

Procedures

1. Connect the circuit as shown in **Figure 32-3A.**

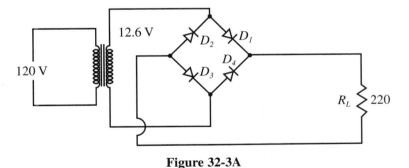

Figure 32-3A

2. Connect the power transformer to ac power.

3. Turn on the power.

4. Using an oscilloscope, draw the waveform of the output across the load resistor R_L in the graph in **Figure 32-3B.**

5. What does the measured waveform indicate about the circuit operation?

_____

_____

_____

(Continued)

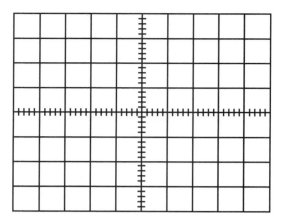

Figure 32-3B

6. What difference do you see between the bridge power supply and the full-wave center-tapped power supply waveforms?

_____

_____

_____

_____

_____

Summing Up

In your notebook, record any observations, problems, and conclusions for this activity.

TRANSISTOR FUNDAMENTALS 33

Activity 33-1:
Transistor Switch

Name _____

Date _____

Class _____ Score _____

Discussion

Transistors have important applications as electronic switches. By receiving an electronic signal, a transistor is capable of opening and closing circuits at very high operating speeds. Transistors are used in switching on and off relays and other control devices.

Objective

In this activity, you will investigate the operation of a transistor switch.

Materials and Equipment

1–560 Ω resistor
1–4.7K Ω resistor
1–2N3904 transistor
1–Breadboard
1–Dc power supply
1–LED
1–VOM

Procedures

1. Connect the circuit as shown in **Figure 33-1A.**

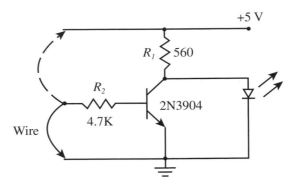

Figure 33-1A

(Continued)

2. Connect the wire from the base resistor to ground.

3. Turn on the power.

4. With reference to ground, measure the base voltage V_B and collector voltage V_C.

5. Record the measurements in the table in **Figure 33-1B.** Record the condition of the LED also.

Wire	V_B	V_C	I_C Cal.	LED (On/off)
GND				
+5 V				

Figure 33-1B

6. Calculate the collector current for each condition of the LED.

7. When the LED is on, is the transistor on or off? _____

8. What would be the result if a square wave were applied to the base resistor?

_____

_____

9. In what way is the application of a square wave to the base like the wire going from ground to 5 V?

_____

_____

_____

_____

Summing Up

In your notebook, record any observations, problems, and conclusions for this activity.

Activity 33-2:	Name _____
Base and Collector	Date _____
Current Relationships	Class _____ Score _____

Discussion

Since the base current controls the collector current, a small change in the base causes a large change in the collector current. This is referred to as the current gain or beta of the transistor. This activity investigates the relationship of the base and collector currents and measures the current gain of a transistor amplifier. Then, it measures the voltage gain of a transistor amplifier.

Objective

In this activity, you will investigate the operation of a transistor amplifier and the relationship of the base and collector currents.

Materials and Equipment

One each of the following resistors:
  2.2K Ω
  12K Ω
  68K Ω
2–0.1 μF capacitors
1–500K potentiometer
1–2N3904 transistor
1–Breadboard
1–Dc power supply
1–Microammeter
1–Milliameter
1–VOM

Procedures

Part 1: Base Current vs. Collector Current

1. Connect the circuit as shown in **Figure 33-2A.**

Caution! Use correct meter polarity.

2. Adjust the potentiometer R_1 for maximum resistance.

3. Turn on the power. Adjust R_1 for an indication of the I_B and I_C currents. Make sure the circuit is operational. Return R_1 to its maximum resistance position.

4. Adjust R_1 for 5 μA and record the collector current I_C in the table in **Figure 33-2B.**

5. Adjust R_1 for each of the other base currents in the table and record the collector current. Since more data may be needed, do not tear down the circuit.

6. Using 1/10 division graph paper, make a graph of the $\dfrac{I_C}{I_B}$ data.

Part 2: Amplifier Voltage Gain

1. Referring to the $\dfrac{I_C}{I_B}$ graph, adjust R_1 so the base current is in the middle of the linear portion of the graph.

2. Apply an input signal as shown in **Figure 33-2C.**

3. While observing the output on an oscilloscope, adjust the amplitude of the input signal so the output is not distorted.

(Continued)

4. Measure the peak-to-peak input and output signal voltages, and record them in the table in **Figure 33-2D.**

5. Calculate the voltage amplification of the amplifier.

$A_V = $ _____

6. In what way is the voltage gain different from the current gain?

_____

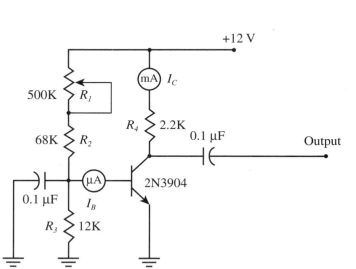

Figure 33-2A

I_B	I_C
0	
5	
10	
15	
20	
25	
30	
35	
40	
45	

Figure 33-2B

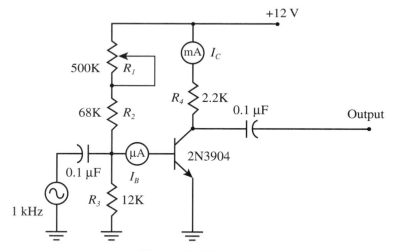

Figure 33-2C

	V_{P-P}	Waveform
V_{IN}		
V_{OUT}		

Figure 33-2D

Summing Up

In your notebook, record any observations, problems, and conclusions for this activity.

Activity 33-3:
Testing Transistors

Name _____

Date _____

Class _____ Score _____

Discussion

Testing a transistor is a common task for a technician. While treating the transistor as two diodes back to back, the simplest test uses an ohmmeter to check for a short across the junctions of the transistor. With the base used as the common connection, each of the junctions (e-b and c-b) are tested for low resistance in one direction and high resistance in the other. A zero resistance indicates a short at that particular junction.

Objective

In this activity, you will test transistors for good, open, or shorted condition.

Materials and Equipment

Set of transistors to be tested (see instructor)
1–VOM

Procedures

1. Ask your instructor for a set of transistors to test.

2. Using an ohmmeter or the diode test position on a digital meter, test each of the transistors for good, open, or shorted condition. Record each of the results in the table in **Figure 33-3.**

3. In the table, note which junction failed, if possible.

Transistor	NPN or PNP	Good	Shorted or Open	Junction Failed
2N 3906				
2N 3055				
TIP 32C				

Figure 33-3

Summing Up

In your notebook, record any observations, problems, and conclusions for this activity.

TRANSISTOR AMPLIFIERS  34

Activity 34-1:
Common Emitter
Amplifier

Name _____

Date _____

Class _____ Score _____

Discussion

The common emitter amplifier is the most frequently used type of transistor amplifier. Although the current gain can vary widely from one transistor to another of the same type, feedback biasing can make the circuit independent of the beta. In this activity, static and dynamic measurements are taken, and the voltage gain of an unbypassed amplifier is compared to one that is bypassed. In all cases, the output waveform should be a sine wave without distortion.

Objective

In this activity, you will investigate a common emitter transistor amplifier.

Materials and Equipment

One each of the following resistors:
   470 Ω
   4.7K Ω
   12K Ω
   100K Ω
2–0.1 µF capacitor
1–2N3904 transistor
1–Audio generator
1–Breadboard
1–Dc power supply
1–Oscilloscope
1–VOM

Procedures

1. Connect the circuit as shown in **Figure 34-1A.**

2. Measure the static dc measurements. Record them in the table in **Figure 34-1B** in the row "Without C_3."

3. Apply a 1 kHz sine wave signal to input capacitor C_1. Adjust the amplitude for no distortion at the output.

(Continued)

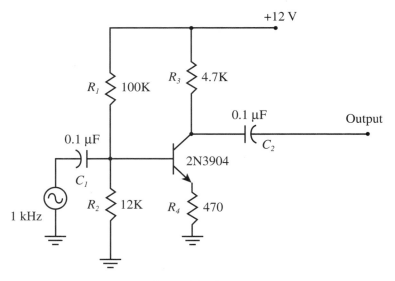

Figure 34-1A

	Static Measurements (No input signal)						
	V_B	V_E	V_C	I_C Calculated	Input P–P	Output P–P	A_V
Without C_3							
With C_3							

Figure 34-1B

4. Measure the input and output signal amplitudes, and calculate the voltage gain A_V of the amplifier. Record them in the table.

5. Add the emitter bypass capacitor C_3 to the circuit as shown in **Figure 34-1C.** Remove the generator from the input and ground the input capacitor.

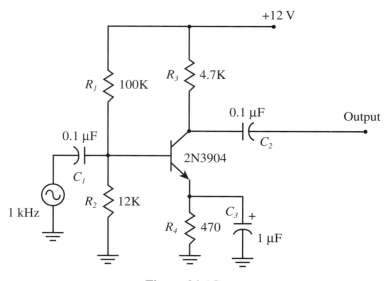

Figure 34-1C

(Continued)

Name _____

6. Take the static dc measurements and record them in the table in the row "With C_3."

7. Apply a 1 kHz sine wave signal to input capacitor C_1. Adjust the amplitude for no distortion at the output.

8. Measure the input and output signal amplitudes and calculate the voltage gain A_V of the amplifier. Record them in the table.

9. What differences occurred in the dc static measurements with and without the emitter bypass capacitor?

   _____

   _____

10. In what way did the emitter bypass capacitor change the voltage gain of the amplifier? Why?

   _____

   _____

   _____

   _____

Summing Up

In your notebook, record any observations, problems, and conclusions for this activity.

Activity 34-2:
Common Collector
Amplifier

Name _____

Date _____

Class _____ Score _____

Discussion

A common collector amplifier is easily identified because its collector is connected directly to the dc source and capacitively bypassed to common ground. The output of a common collector amplifier is taken from the emitter. The resistor network on the base forms a voltage divider that usually causes half the source to be applied to the base.

For some circuitry, a better Q-point can be obtained if the base resistance is increased until the emitter voltage is equal to half the collector voltage. This also makes possible the use of direct coupling with a common emitter circuit as a driver circuit to the input of the common collector amplifier.

Objective

In this activity, you will investigate a common collector (emitter follower) transistor amplifier.

Materials and Equipment

One each of the following resistors:
  2K Ω
  100K Ω
  120K Ω
2–0.1 µF capacitors
1–0.22 µF capacitor
1–2N2222 transistor
1–Audio generator
1–Breadboard
1–Dc power supply
1–Oscilloscope
1–VOM

Procedures

1. Connect the circuit as shown in **Figure 34-2A.**

2. Measure the static dc measurements and record them in the table in **Figure 34-2B.**

3. Apply a 1 kHz sine wave signal to input capacitor C_1. Adjust the amplitude for no distortion at the output.

4. Measure the input and output signal amplitudes, and calculate the voltage gain A_V of the amplifier. Record these in the table.

5. How does the common collector amplifier compare to the common emitter amplifier in Activity 34-1?

_____

_____

_____

(Continued)

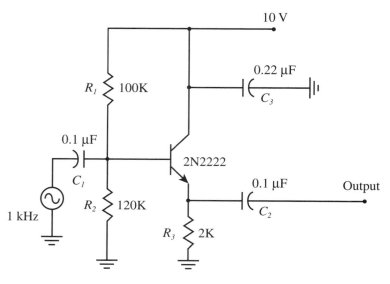

Figure 34-2A

Static Measurements (No input signal)			Input P–P	Output P–P	A_V
V_B	V_E	V_C			

Figure 34-2B

Summing Up

In your notebook, record any observations, problems, and conclusions for this activity.

Activity 34-3:	Name _____
Voltage Divider	Date _____
Bias Effects	Class _____ Score _____

Discussion

An often used method of biasing a common emitter amplifier is to connect the base to a voltage divider across the source. Although a carefully calculated bias network can be a place to start, the actual resistors used are selected by experimentally substituting resistors until the right combination is found. This activity investigates the changes in biasing as the bias network is changed.

Objective

In this activity, you will investigate the effects of changing the voltage divider bias in a common emitter transistor amplifier.

Materials and Equipment

One each of the following resistors:
  470 Ω
  4.7K Ω
  47K Ω
  100K Ω
  220K Ω
  1 Meg Ω
2–10K Ω resistors
2–0.1 μF capacitors
1–2N3904 transistor
1–Audio generator
1–Breadboard
1–Dc power supply
1–Oscilloscope
1–VOM

Procedures

1. Connect the circuit as shown in **Figure 34-3A** using the 100K Ω resistor for R_1.

2. Measure the static dc measurements and record them in the table in **Figure 34-3B.**

3. Apply a 1 kHz sine wave signal to input capacitor C_1 and the amplitude to a 20 mV RMS input signal.

4. Measure the input and output signal amplitudes and calculate the voltage gain A_V of the amplifier. Record these in the table. The output signal should be a sine wave. Record this in the table also.

5. Without changing any other values or adjustments, replace R_1 with the next value in the table.

6. Measure the input and output signal amplitudes and calculate the voltage gain A_V of the amplifier. Record these in the table. The output signal may not be a sine wave. Record this in table also.

(Continued)

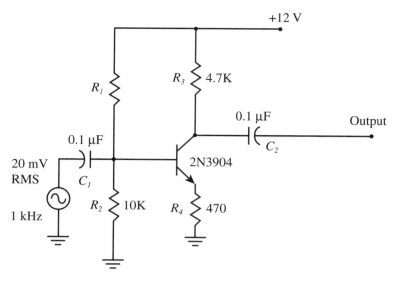

Figure 34-3A

R_1	V_B	V_E	V_C	V_O	A_V	Waveform
100K						
220K						
47K						
1 Meg						
10K						

Figure 34-3B

7. Complete the table by repeating steps 5 and 6 with the other values of R_1 in the table.

8. What happened when R_1 became 1 megohm?

_____

_____

_____

Summing Up ─────────────────────────────────

In your notebook, record any observations, problems, and conclusions for this activity.

Activity 34-4: Two-stage Amplifier

Name _____

Date _____

Class _____ Score _____

Discussion

A single amplifier stage is not very helpful for a complete amplifying task. Two or more amplifier stages are coupled together to give higher voltage amplification. The design of multiple-stage amplifiers is not simple. Engineering skills in solid-state circuit design are required. This activity investigates the operation of a two-stage voltage amplifier.

Objective

In this activity, you will investigate the operation of a two-stage voltage amplifier.

Materials and Equipment

One each of the following resistors:
   270 Ω
   390 Ω
   3.3K Ω
   4.7K Ω
Two each of the following resistors:
   12K Ω
   100K Ω
3–0.1 µF capacitors
2–2N3904 transistors
1–Audio generator
1–Breadboard
1–Dc power supply
1–Oscilloscope
1–VOM

Procedures

1. Connect the circuit as shown in **Figure 34-4A.**

2. Measure the static dc measurements of Q_1 and Q_2 and record them in the table in **Figure 34-4B.**

3. Apply a 1 kHz sine wave signal to input capacitor C_1. Adjust the input signal amplitude for no distortion of the output signal.

4. Measure the input and output signal amplitudes of Q_1 and Q_2. Calculate the voltage gain A_V of each amplifier and the overall A_V of the two stages. Record these in the table.

5. Using the overall voltage gain ratio, calculate the gain in decibels and record it here.

   Overall A_V = _____ dB = _____

6. Connect an 8 Ω speaker from the output capacitor to ground.

7. What happens to the output signal when the speaker is connected?

_____

(Continued)

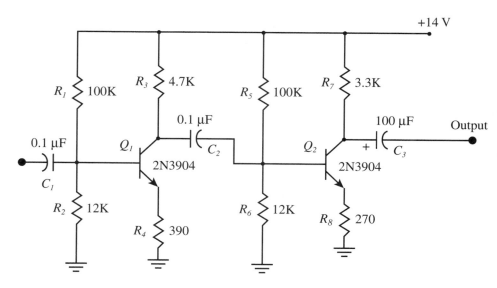

Figure 34-4A

Q_1

	Measured				Calculated	
V_B	V_E	V_C	V_{IN}	V_{OUT}	I_C	A_V

Q_2

	Measured				Calculated	
V_B	V_E	V_C	V_{IN}	V_{OUT}	I_C	A_V

Figure 34-4B

8. Measure the peak-to-peak voltage across the speaker.

    $V_{P-P} =$ _____

9. Calculate the output power. Remember to change the peak-to-peak voltage to RMS before calculating.

    $P = \dfrac{E}{R}$

    $P =$ _____

Note: If possible, save this circuit to compare loudness with the next activity.

Summing Up

In your notebook, record any observations, problems, and conclusions for this activity.

Activity 34-5:
Power Amplifier

Name _____

Date _____

Class _____ Score _____

Discussion

Although a two-stage amplifier increases the voltage gain, power is required to drive such things as speakers. A power amplifier from several milliwatts to several hundred watts is needed. The complementary-symmetry power amplifier has the advantage of not requiring a transformer, making it lightweight and inexpensive to manufacture.

Objective

In this activity, you will investigate the operation of a complementary-symmetry power amplifier.

Materials and Equipment

1–390 Ω resistor
2–1.5K Ω resistors
1–4.7K Ω resistor
2–10K Ω resistors
1–12K Ω resistor
2–2N3904 transistors
1–2N3906 transistor
2–0.1 µF capacitors
1–100 µF capacitor
1–Audio generator
1–Breadboard
1–Dc power supply
1–Oscilloscope
1–8 Ω speaker
1–VOM

Procedures

1. Connect the circuit as shown in **Figure 34-5.**

2. Apply a 1 kHz sine wave signal to input capacitor C_1, and adjust the input signal amplitude for no distortion of the output signal.

3. Measure the input and output signal amplitudes.

   $V_{IN} = $ _____

   $V_{OUT} = $ _____

4. Calculate the voltage gain.

   $A_V = $ _____

5. Calculate the gain in decibels.

   $dB = $ _____

6. Connect an 8 Ω speaker from the output capacitor to ground.

(Continued)

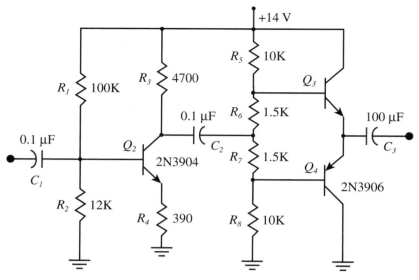

Figure 34-5

7. Measure the peak-to-peak voltage across the speaker.

$V_{P-P} =$ _____

8. Calculate the output power. Remember to change the peak-to-peak voltage to RMS.

$$P = \frac{E^2}{R}$$

$P =$ _____

9. How does the complementary-symmetry power amplifier compare in loudness with the two-stage amplifier in Activity 34-3?

_____

_____

_____

Summing Up

In your notebook, record any observations, problems, and conclusions for this activity.

MISCELLANEOUS DEVICES  35

Activity 35-1:
SCR Motor
Control Circuit

Name _____

Date _____

Class _____ Score _____

Discussion

An SCR is commonly used as a latching switch in dc circuits, such as in motors and alarms. Once triggered, the SCR remains in the on state until the current through the SCR is stopped. Since the gate current affects the breakover voltage of the SCR, the larger the gate current, the lower the anode-to-cathode voltage at which the forward breakover voltage will occur. In normal designs, the gate triggering signal is large enough to guarantee complete SCR turn-on.

Objective

In this activity, you will investigate an SCR control of a dc motor circuit.

Materials and Equipment

One each of the following resistors:
  1K Ω
  3.3K Ω
  10K Ω
1–6 V to 12 V dc motor
1–2N3906 transistor
1–Oscilloscope
2–SPST switches
1–C106B SCR
1–VOM

Procedures

1. Connect the circuit as shown in **Figure 35-1A.**

2. Measure the transistor collector and the SCR gate voltage. Record these in the "Motor off" row of the table in **Figure 35-1B.**

3. Momentarily close switch S_1. What action takes place?

_____

4. Again measure the transistor collector and the SCR gate voltage. Record them in the "Motor on" row of the table.

5. Momentarily open switch S_2. What action takes place? Why?

_____

_____

_____

_____

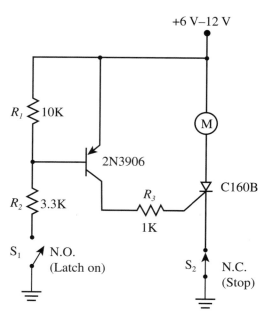

Figure 35-1A

	V_B	V_C	V_{GATE}
Motor off			
Motor on			

Figure 35-1B

Summing Up

In your notebook, record any observations, problems, and conclusions for this activity.

Activity 35-2:
DIAC Trigger Voltage

Name _____

Date _____

Class _____ Score _____

Discussion

When the breakover voltage of a DIAC is exceeded, the voltage across the DIAC drops immediately. DIACs are manufactured with breakover voltage from 10 V to 40 V. The amount of current flowing through the DIAC must not exceed the power dissipation of the DIAC. If the power dissipation is 250 mW and the breakover voltage is 20 V, the current is:

$$I = \frac{P}{V}$$

$$= \frac{250 \times 10^{-3}}{20}$$

$$= 12.5 \text{ mA}$$

This is the absolute maximum current for the DIAC. The minimum value of the limiting resistor can be calculated:

$$R = \frac{V}{I}$$

$$= \frac{20}{12.5 \times 10^{-3}}$$

This value would make the DIAC operate at maximum power dissipation. To guarantee a long DIAC working life, this resistance should be increased to twice the calculated value. The resistance is then 3200 Ω, and the closest standard value resistor for the circuit should be used.

Objective

In this activity, you will measure the trigger voltage of a DIAC.

Materials and Equipment

1–4.7K Ω resistor
1–100K potentiometer
1–HS-10-ND or similar DIAC
1–30 V to 50 V power supply
1–Breadboard
1–VOM

Procedures

1. Connect the circuit as shown in **Figure 35-2.**

2. Connect the voltmeter to the circuit so your hands are free.

3. Adjust the potentiometer to the ground position so 0 V are placed on the DIAC.

4. While observing the voltmeter, slowly adjust the potentiometer. The meter should indicate a rise in voltage. Continue to increase the voltage until, at a certain voltage value, the voltage suddenly drops. This is the breakover voltage.

$V_{TRIG} =$ _____

(Continued)

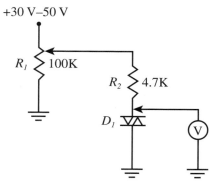

Figure 35-2

5. What is the purpose of resistor R_2?

_____

6. What happens if resistor R_2 is made smaller in value?

_____

_____

Summing Up

    In your notebook, record any observations, problems, and conclusions for this activity.

Activity 35-3: Testing LEDs

Name _____

Date _____

Class _____ Score _____

Discussion

Placing a component in a known circuit provides one of the best tests for that component. In this activity, the circuit can be used to test one or several LEDs or to test the individual LEDs in a seven-segment LED display. Resistor R_1 limits the current through the LED.

Objective

In this activity, you will test an LED for proper operation.

Materials and Equipment

1–560 Ω resistor
1–Breadboard
1–Dc power supply
1–LED
1–VOM

Procedures

1. Connect the circuit as shown in **Figure 35-3.** Make sure the cathode of the LED is in the correct position. The LED will light up if it is good.

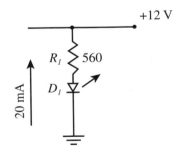

Figure 35-3

2. What value should the limiting resistor be for a 24 V supply? _____

   For a 5 V supply? _____

3. What would happen if the limiting resistor were too small in value?

   _____

Summing Up

In your notebook, record any observations, problems, and conclusions for this activity.

Activity 35-4:
TRIAC Circuits

Name _____

Date _____

Class _____ Score _____

Discussion

Since TRIACs conduct current in both directions, they are used in ac motor control circuits and to replace mechanical relays. This makes circuits smaller, lighter in weight, and more reliable. Unlike the SCR, there is no need to stop the current to turn off the TRIAC because it turns off automatically when the ac current changes direction. Though unnecessary for circuit operation, capacitors and a DIAC are used in TRIAC circuits to gain better control of the load. The capacitors extend the triggering angle beyond 90° and allow less current to flow in the load. This activity uses a simple light dimmer circuit to demonstrate the TRIAC.

Objective

In this activity, you will demonstrate the operation of a TRIAC in an alternating current circuit.

Materials and Equipment

One each of the following resistors:
   10K Ω
   560K Ω
2–0.1 µF 500 V dc capacitors
1–0.01 µF 500 V dc capacitor
1–500K potentiometer
1–DIAC, HS-10-ND or equivalent
1–TRIAC, SC146M or equivalent
1–Breadboard
1–Table lamp or similar load
1–VOM

Procedures

Caution! 120 V ac circuits are dangerous.

1. Connect the circuit as shown in **Figure 35-4.**

2. Adjust potentiometer R_2 to the center of its range.

3. Connect the load to the circuit.

4. Connect the circuit to a 120 V ac source.

5. Adjust the potentiometer R_2 resistance up and down and observe the results.

6. What is the purpose of resistor R_2?

_____

7. Turn the potentiometer to the OFF position.

8. Connect an ac voltmeter across the DIAC.

(Continued)

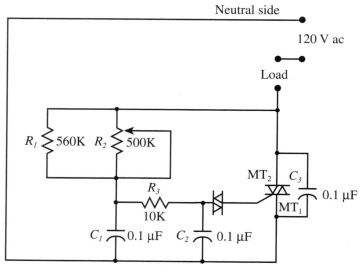

Figure 35-4

9. Slowly turn potentiometer R_2 and observe the voltmeter. At a certain voltage, the voltage drop across the DIAC will drop to its forward operating voltage and the lamp will turn on.

   $V_{ON} = $ _____

10. Continue to increase the intensity of the lamp and observe the voltage drop across the DIAC. What is the voltage drop across the DIAC when the lamp is fully on?

    $V = $ _____

11. What would the result be if potentiometer R_2 were made smaller in value?

    _____

    _____

Summing Up

In your notebook, record any observations, problems, and conclusions for this activity.

Activity 35-5:
Opto-isolator Circuits

Name _____

Date _____

Class _____ Score _____

Discussion

Completely free from the influence of light from the outside world, opto-isolators or optocouplers connect circuits together using only the light within the device. No actual electrical connection is made between the input and output. The output is always a faithful reproduction of the input with some loss of signal voltage or gain. Since the output is not connected electrically to the input, any connection across the output is not reflected back into the input circuits. In this activity, the opto-isolator is connected to a simple amplifier to demonstrate its ability to pass a signal.

Objective

In this activity, you will demonstrate the operation of an opto-isolator circuit.

Materials and Equipment

One each of the following resistors:
  220 Ω
  680 Ω
  1K Ω
  2.2K Ω
  12K Ω
  100K Ω
3–0.1 μF capacitors
1–4N28 opto-isolator
1–2N3904 transistor
1–Oscilloscope
1–VOM

Procedures

1. Connect the circuit as shown in **Figure 35-5.**

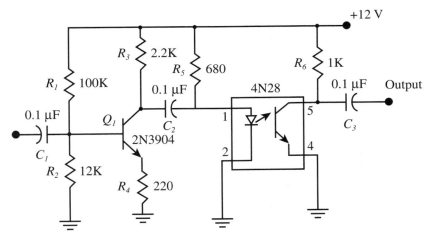

Figure 35-5

(Continued)

2. Set the input signal voltage for 1 V peak-to-peak.

3. Measure the gain of transistor Q_1.

$Q_1 V_{OUT} = $ _____

$A_V = $ _____

4. Measure the input and output signals of the opto-isolator.

$V_{IN} = $ _____

$V_{OUT} = $ _____

5. Calculate the gain of the opto-isolator.

$A_V = $ _____

6. What advantage does the opto-isolator have compared to other circuits?

_____

_____

7. What disadvantage does the opto-isolator have compared to other circuits?

_____

_____

Summing Up

In your notebook, record any observations, problems, and conclusions for this activity.

INTEGRATED CIRCUITS ► 36

Activity 36-1:
Inverting and Noninverting
Operational Amplifiers

Name _____

Date _____

Class _____ Score _____

Discussion

One common circuit component of operational amplifiers is feedback resistor R_F connected from the output to the input. By controlling the amount of feedback, this resistor controls the gain of the op-amp. The same type op-amp device can be used in the design of many types of circuits. In addition, op-amps have inverting and noninverting signal input terminals, making them even more versatile in circuit design.

Objective

In this activity, you will investigate the operation of inverting and noninverting operational amplifiers.

Materials and Equipment

One each of the following resistors:
    1K Ω
    10K Ω
    33K Ω
    56K Ω
    100K Ω
    1 Meg Ω
1–0.1 µF capacitor
1–1 µF electrolytic or tantalum capacitor
1–741 op-amp
1–Audio generator
1–Breadboard
1–Dual polarity dc power supply
1–Oscilloscope
1–VOM

Procedures

Part 1: Amplifier Gain

1. Connect the circuit as shown in **Figure 36-1A.**

(Continued)

211

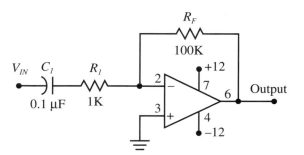

Figure 36-1A

2. Apply a 1 kHz sine wave signal to input capacitor C_1. Adjust the input signal amplitude for no distortion of the output signal.

   $V_{IN} =$ _____

3. Measure the output signal amplitude and record it in the table in **Figure 36-1B.**

R_F	V_O	A_V
10K		
56K		
100K		
1 Meg		

Figure 36-1B

4. Change resistor R_F to 10K Ω in the table and repeat step 3.

5. Change resistor R_F to each of the remaining resistors in the table and repeat step 3.

6. Calculate the voltage gain A_V for each value of R_F and record them in the table.

7. What is the relationship of resistor R_F to the gain of the amplifier?

   _____

8. What difference was seen in the 1 megohm resistor compared to the other resistors?

   _____

   _____

Part 2: Input-Output Phase Relationship

1. Using a dual trace oscilloscope, connect channel 1 to the input and channel 2 to the output, as shown in **Figure 36-1C.**

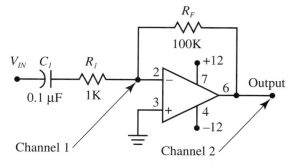

Figure 36-1C

(Continued)

Name _____

2. Apply a 1 kHz sine wave signal to input capacitor C_1. Adjust the input signal amplitude for no distortion of the output signal.

3. What is the phase relationship of the input and output waveforms?

_____

_____

Part 3: Noninverting Amplifier Phase Relationship

1. Using a dual trace oscilloscope, connect channel 1 to the input and channel 2 to the output, as shown in **Figure 36-1D.**

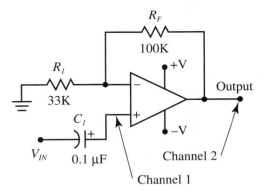

Figure 36-1D

2. Apply a 1 kHz sine wave signal to input capacitor C_1. Adjust the input signal amplitude for no distortion of the output signal.

3. What is the phase relationship of the input and output waveforms?

_____

_____

4. How does the inverting amplifier compare to the noninverting amplifier?

_____

_____

Summing Up ━━━━━━━━━━━━━━━━━━━━━━━━━━━

    In your notebook, record any observations, problems, and conclusions for this activity.

Activity 36-2:
Dc Offset Voltage
Adjustment

Name _____

Date _____

Class _____ Score _____

Discussion

Mismatched parts of the internal components of op-amps cause imbalances in the circuitry. One of those imbalances is caused by the input offset voltage. Many op-amps provide terminals identified on data sheets as offset null or balance terminals. These terminals are internally connected to affect the internal circuitry of the op-amp. Usually a multiturn potentiometer is used so small adjustments of the voltage are possible on the null terminals. For op-amps that do not have null terminals, data sheets indicate a method of connecting a circuit externally to null the op-amp.

Objective

In this activity, you will demonstrate the dc offset adjustment of an op-amp.

Materials and Equipment

One each of the following resistors:
  1K Ω
  100K Ω
1–0.1 μF capacitor
1–741 op-amp
1–10K potentiometer
1–Audio generator
1–Breadboard
1–Dual polarity dc power supply
1–Oscilloscope
1–VOM

Procedures

1. Connect the circuit as shown in **Figure 36-2.**

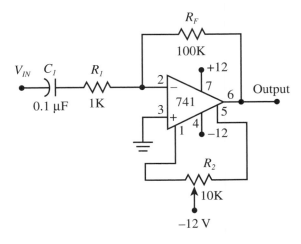

Figure 36-2

(Continued)

2. Temporarily connect the input V_{IN} to ground and connect a voltmeter to the output. At this time the output voltage is:

   $V_{OUT} = $ _____

3. Adjust the offset balance (or null) potentiometer R_2 for zero output.

4. Apply a 1 kHz sine wave signal to input capacitor C_1. Adjust the input signal amplitude for no distortion of the output signal.

   $V_{IN} = $ _____

5. Measure the output signal amplitude.

   $V_{OUT} = $ _____

6. What is the voltage gain of the amplifier?

   $A_V = $ _____

7. What effect does the offset have on the input-output phase relationship of the amplifier?

   _____

Summing Up

In your notebook, record any observations, problems, and conclusions for this activity.

Activity 36-3:	Name _____
Integrated Circuit	Date _____
Amplifiers	Class _____ Score _____

Discussion

Integrated circuits are used extensively in analog signal applications. Some uses of analog integrated circuits are audio amplifiers, RF amplifiers, and operational amplifiers. This activity uses a small power amplifier to demonstrate integrated circuit amplifiers. Although a heat sink is not required with this integrated circuit, never operate an integrated circuit unmounted from its heat sink. Excessive heat will damage the device. When working with integrated circuits, make sure the index mark is in the correct position and the power supply is of the correct voltage, polarity, and pin position.

Objective

In this activity, you will demonstrate an integrated circuit amplifier.

Materials and Equipment

One each of the following resistors:
  10 Ω
  15K Ω
2–0.047 µF capacitors
1–0.1 µF capacitor
1–100 µF electrolytic capacitor
1–386 IC power amplifier
1–Audio generator
1–Breadboard
1–Dc power supply
1–Oscilloscope
1–8 Ω speaker
1–VOM

Procedures

1. Connect the circuit as shown in **Figure 36-3A.**

2. Apply a 1 kHz sine wave signal to input capacitor C_1. Adjust the input signal amplitude for no distortion of the output signal.

3. Measure the input and output signal amplitudes.

   V_{IN} = _____

   V_{OUT} = _____

4. What is the voltage gain of the amplifier?

   A_V = _____

5. Calculate the gain of the amplifier in decibels.

   dB = _____

(Continued)

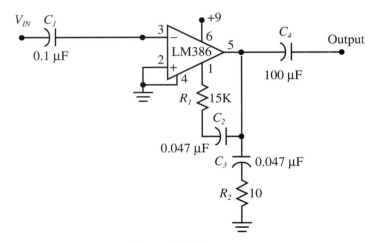

Figure 36-3A

6. Connect a capacitor and speaker to the output terminal, pin 5, as shown in **Figure 36-3B.**

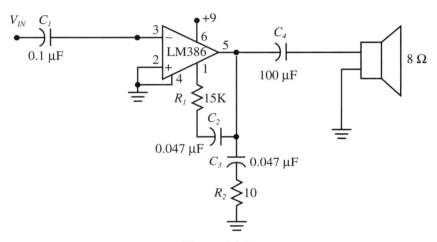

Figure 36-3B

7. Turn the power on, and adjust the generator output for maximum undistorted output. The speaker may cause some reflected distortion.

8. Measure the output voltage across the speaker.

   $V_{OUT} = $ _____

9. Calculate the output power of the amplifier.

   $P_{OUT} = $ _____

Summing Up

In your notebook, record any observations, problems, and conclusions for this activity.

Activity 36-4:
Summing Amplifiers

Name _____

Date _____

Class _____ Score _____

Discussion

Electronic circuits are used for mathematical functions. A summing amplifier is used in instrumentation or industrial control systems. The output voltage of the circuit is the sum of the individual input voltages. As stated in the textbook, V_1, V_2, and V_3 are added so the output is:

$$V_O = -R_F \left(\frac{V_1}{R_1} + \frac{V_2}{R_2} + \frac{V_3}{R_3} \right)$$

If feedback resistor R_F and the input resistances are equal (precision), the output voltage is:

$$V_O = -(V_1 + V_2 + V_3)$$

This activity uses a voltage divider to supply various dc voltages to the input for summing. Once this activity is completed, other resistor values can be substituted into the voltage divider, and the summing amplifier actions can be compared.

Objective

In this activity, you will demonstrate the operation of a summing amplifier.

Materials and Equipment

1–100 Ω resistor
2–150 Ω resistors
1–180 Ω resistor
4–100K Ω resistors
1–741 op-amp
1–10K potentiometer
1–Breadboard
1–Dual polarity dc power supply
1–VOM

Procedures

1. Connect the circuit as shown in **Figure 36-4A.**

2. Disconnect R_5, R_6, and R_7 at the disconnect point.

3. Connect the inverting input to ground, and adjust offset adjustment R_9 for 0 V at the output. If 0 V cannot be achieved, adjust R_9 to the point of switching from negative to positive.

4. Calculate the voltages of voltage divider R_1, R_2, R_3, and R_4.

5. Measure the voltages at points V_1, V_2, and V_3.

Calculated	Measured
$V_1 =$ –––	$V_1 =$ –––
$V_2 =$ –––	$V_2 =$ –––
$V_3 =$ –––	$V_3 =$ –––

(Continued)

6. How do the measured values compare to the calculated values?

_____

_____

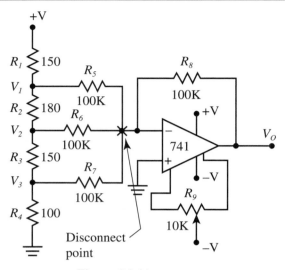

Figure 36-4A

7. With R_5, R_6, and R_7 disconnected, measure V_O and record it in the table in **Figure 36-4B**.

Connect	V_O
None	
R_7	
R_6	
R_5	
R_6, R_7	

Figure 36-4B

8. Reconnect R_7, measure V_O, and record it in the table.

9. Continue with the remaining resistor connections given in the table and record the results.

10. What is the comparison of the measured values of V_1, V_2, and V_3 to V_O in the table?

_____

_____

_____

Summing Up

In your notebook, record any observations, problems, and conclusions for this activity.

Activity 36-5:
555 Timer Circuit

Name _____

Date _____

Class _____ Score _____

Discussion

In the 1970s, a revolutionary type of linear integrated circuit was developed. The 555 timer is an inexpensive, easy-to-use device that can operate in either a monostable or astable mode. In some applications, the stability of the 555 circuit is limited to the stability of the RC components. This activity only touches on the astable timer. Complete textbooks describing the 555 and its applications are available.

Objective

In this activity, you will demonstrate the astable operation of the 555 timer.

Materials and Equipment

2–0.01 µF capacitor
1–50K potentiometer
1–10K Ω resistor
1–555 timer
1–Breadboard
1–Dc power supply
1–Oscilloscope
1–VOM

Procedures

1. Connect the circuit as shown in **Figure 36-5A.**

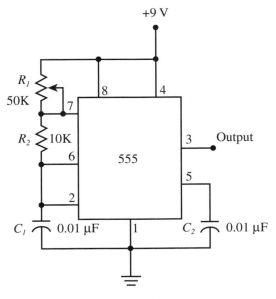

Figure 36-5A

(Continued)

2. Connect an oscilloscope to output pin 3 and observe the output waveform.

3. Turn potentiometer R_1 to the center of its rotation. Turn on the power.

4. Rotate potentiometer R_1 fully clockwise. This adjustment should cause the waveform to flat-line. If it does not, reverse the connections of R_1 on pins 7 and 8 of the integrated circuit.

5. Turn R_1 clockwise until the waveform flat-lines; then, adjust it until the waveform first appears. If any overshoot exists, continue turning the adjustment until it disappears.

6. Measure the frequency of the waveform.

   $f =$ _____

7. What is the duty cycle of the waveform?

   Percent duty cycle = _____

8. Sketch the waveform on the graph in **Figure 36-5B.**

9. Turn R_1 counterclockwise.

10. Measure the frequency of the waveform.

    $f =$ _____

11. What is the duty cycle of the waveform?

    Percent duty cycle = _____

12. Sketch the waveform on the graph in **Figure 36-5C.**

13. Why doesn't the waveform stay symmetrical?

    _____

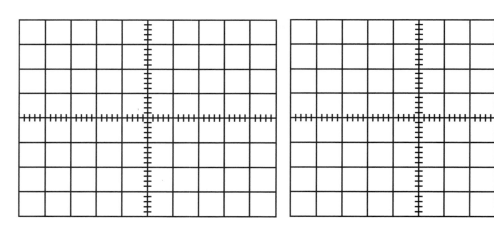

Figure 36-5B   **Figure 36-5C**

Summing Up

In your notebook, record any observations, problems, and conclusions for this activity.

INTRODUCTION TO DIGITAL ELECTRONICS

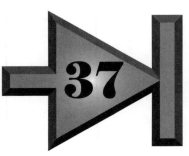

37

Activity 37-1:
Inverters

Name _____

Date _____

Class _____ Score _____

Discussion

An inverter "inverts" the signal that is applied to its input. If the input is a logic 0, the output of the inverter is a logic 1. If two inverters are connected, the output of the first inverter is the input for the second inverter. In this activity you will study the operation of a single inverter, then connect all six inverters of the 7404 integrated circuit.

Objective

In this activity, you will demonstrate the operation of logic inverters.

Materials and Equipment

1–5 V dc power supply
1–7404 inverter IC
6–150 Ω resistors
1–Breadboard
6–LEDs
1–SPDT switch
1–VOM

Procedures

1. Wire the switch so it can be connected to a breadboard.

2. Insert the 7404 IC into the breadboard.

3. With the power off, connect the circuit as shown in **Figure 37-1A.** Refer to **Figure 37-1B** for power and ground connections.

4. Turn the power on. With the input connected to ground for a logic 0 and 5 V for a logic 1, complete the table in **Figure 37-1C.**

5. Turn the power off.

6. What is the relationship of the input and output logic levels?

   _____

   _____

7. Write the Boolean expression for the circuit.

   _____

(Continued)

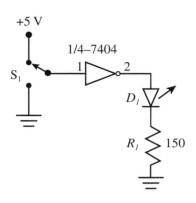

Figure 37-1A

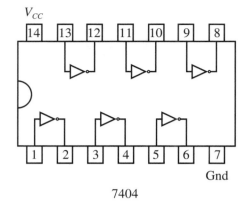

7404

Figure 37-1B

Input	Output
A	$\overline{A}$
0	
1	

Figure 37-1C

8. Connect the circuit as shown in **Figure 37-1D.**

9. Turn the power on. With the input connected to ground for a logic 0 and 5 V for a logic 1, complete the table in **Figure 37-1E.**

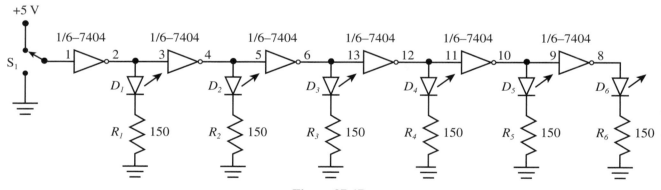

Figure 37-1D

Input	Outputs					
S_1	D_1	D_2	D_3	D_4	D_5	D_6
0						
1						

Figure 37-1E

(Continued)

Name _____

10. Turn the power off.

11. Look over the results in the table and decide if the six inverters are operating correctly. If an LED fails to light, check to see if the cathode is connected to ground.

12. What is the relationship of the input to the output of the last inverter?

_____

_____

Summing Up

In your notebook, record any observations, problems, and conclusions for this activity.

Activity 37-2:
AND and NAND Gates

Name _____

Date _____

Class _____ Score _____

Discussion

Inputs and outputs can be represented by a chosen combination including yes-no, true-false, high-low, 5 V-0 V, A-B and X, or any symbol combination. In all cases, the combinations represent a logic 0 or logic 1. For the AND gate to have output, both inputs must be at a logic 1, while the NAND gate output is just the opposite.

Objective

In this activity, you will demonstrate the operation of logic AND and NAND gates.

Materials and Equipment

1–5 V dc power supply
1–7408 AND IC
1–7404 inverter IC
1–7400 NAND IC
1–150 Ω resistor
1–Breadboard
2–LEDs
3–SPDT switches
1–VOM

Procedures

Part 1: AND Gates

1. Wire the switches so they can be connected to the breadboard.

2. Insert the 7408 IC into a breadboard.

3. With the power off, connect the circuit as shown in **Figure 37-2A.** Refer to **Figure 37-2B** for power and ground connections.

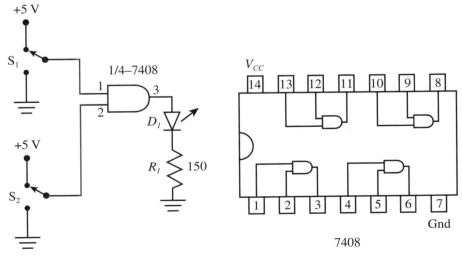

Figure 37-2A

Figure 37-2B

(Continued)

4. Turn the power on. With the input connected to ground for a logic 0 and 5 V for a logic 1, complete the table in **Figure 37-2C.**

Inputs		Output
S_2	S_1	D_1
0	0	
0	1	
1	0	
1	1	

Figure 37-2C

5. Turn the power off.

6. What is the relationship of the input and output logic levels?

_____

_____

7. Write the Boolean expression for the circuit.

_____

_____

8. Connect the circuit as shown in **Figure 37-2D.**

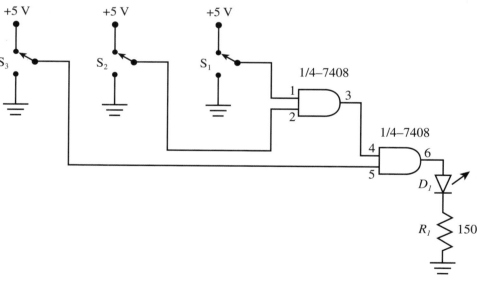

Figure 37-2D

9. Turn the power on. With the input connected to ground for a logic 0 and 5 V for a logic 1, complete the table in **Figure 37-2E.**

10. Turn the power off.

11. Look over the results in the table and decide if the circuit is operating correctly.

(Continued)

Name _____

	Inputs		Output
S_2	S_2	S_1	D_1
0	0	0	
0	0	1	
0	1	0	
0	1	1	
1	0	0	
1	0	1	
1	1	0	
1	1	1	

Figure 37-2E

12. What is the relationship of the input to the output of the circuit?

_____

13. Write the Boolean expression for the circuit.

_____

_____

Part 2: NAND Gates

1. Insert the 7408 and 7404 ICs into a breadboard.

2. With the power off, connect the circuit as shown in **Figure 37-2F.** Refer to Figures 37-1B and 37-2B for power and ground connections.

3. Turn the power on. With the input connected to ground for a logic 0 and 5 V for a logic 1, complete the table in **Figure 37-2G.**

4. Turn the power off.

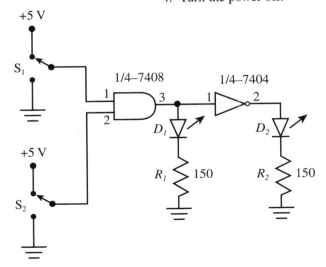

Figure 37-2F

Inputs		Outputs	
S_2	S_1	D_1	D_2
0	0		
0	1		
1	0		
1	1		

Figure 37-2G

(Continued)

5. What is the relationship of the input and output logic levels?

6. Write the Boolean expression for the circuit.

7. Connect the circuit as shown in **Figure 37-2H.** Refer to **Figure 37-2I** for power and ground connections.

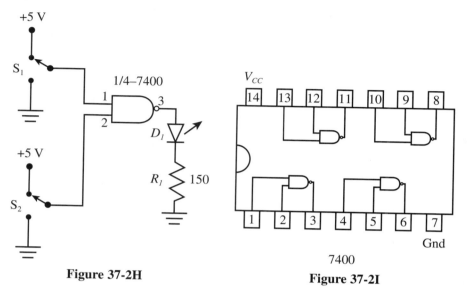

Figure 37-2H   **Figure 37-2I**

8. Turn the power on. With the input connected to ground for a logic 0 and 5 V for a logic 1, complete the table in **Figure 37-2J.**

Inputs		Output
S_2	S_1	D_1
0	0	
0	1	
1	0	
1	1	

Figure 37-2J

9. Turn the power off.

10. Look over the results in the table and decide if the circuit is operating correctly.

11. What is the relationship of the input to the output of the circuit?

12. Write the Boolean expression for the circuit.

Summing Up

In your notebook, record any observations, problems, and conclusions for this activity.

Activity 37-3:	Name
OR and NOR Gates	Date
	Class _____ Score _____

Discussion

The OR gate, sometimes referred to as the "any or all gate," will produce an output when there is input at any one of the inputs. The OR gate may also be described as the inclusive OR gate because the output is dependent on all the inputs. The action of all the inputs is included in the output.

Objective

In this activity, you will demonstrate the operation of logic OR and NOR gates.

Materials and Equipment

1–5 V Dc power supply
1–7432 OR IC
1–7402 NOR IC
1–150 Ω resistor
1–Breadboard
1–LED
4–SPDT switches
1–VOM

Procedures

Part 1: OR Gates

1. Wire the switches so they can be connected to a breadboard.

2. Insert the 7432 IC into the breadboard.

3. With the power off, connect the circuit as shown in **Figure 37-3A.** Refer to **Figure 37-3B** for power and ground connections.

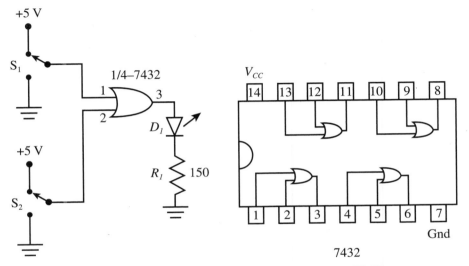

Figure 37-3A   Figure 37-3B

(Continued)

4. Turn the power on. With the input connected to ground for a logic 0 and 5 V for a logic 1, complete the table in **Figure 37-3C.**

Inputs		Output
S_2	S_1	D_1
0	0	
0	1	
1	0	
1	1	

Figure 37-3C

5. Turn the power off.

6. What is the relationship of the input and output logic levels?

_____

_____

7. Write the Boolean expression for the circuit.

_____

8. Connect the circuit as shown in **Figure 37-3D.**

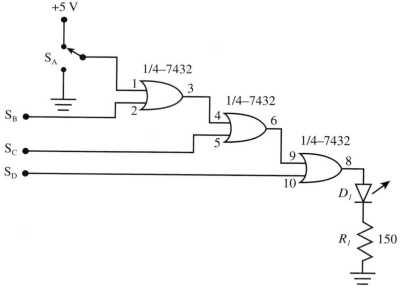

Figure 37-3D

9. Turn the power on. With the input connected to ground for a logic 0 and 5 V for a logic 1, complete the table in **Figure 37-3E.**

10. Turn the power off.

11. Look over the results in the table and decide if the circuit is operating correctly.

(Continued)

Name _____

Inputs				Output
S_D	S_C	S_B	S_A	D_1
0	0	0	0	
0	0	0	1	
0	0	1	0	
0	0	1	1	
0	1	0	0	
0	1	0	1	
0	1	1	0	
0	1	1	1	
1	0	0	0	
1	0	0	1	
1	0	1	0	
1	0	1	1	
1	1	0	0	
1	1	0	1	
1	1	1	0	
1	1	1	1	

Figure 37-3E

12. What is the relationship of the input to the output of the circuit?

_____

13. Write the Boolean expression for the circuit.

_____

_____

Part 2: NOR Gates

1. Insert the 7402 into a breadboard.

2. With the power off, connect the circuit as shown in **Figure 37-3F.** Refer to **Figure 37-3G** for power and ground connections.

3. Turn the power on. With the input connected to ground for a logic 0 and 5 V for a logic 1, complete the table in **Figure 37-3H.**

4. Turn the power off.

5. What is the relationship of the input and output logic levels?

_____

(Continued)

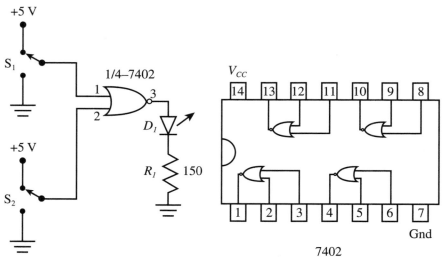

Figure 37-3F   **Figure 37-3G**

Inputs		Output
S_2	S_1	D_1
0	0	
0	1	
1	0	
1	1	

Figure 37-3H

6. Write the Boolean expression for the circuit.

_____

_____

Summing Up

   In your notebook, record any observations, problems, and conclusions for this activity.

Activity 37-4:
Exclusive OR and
XNOR Gates

Name _____

Date _____

Class _____ Score _____

Discussion

The exclusive OR gate will produce an output when there is input at either of, but not both of, the inputs. The zero-zero and one-one input combinations are excluded from having output. The exclusive OR gate is sometimes referred to as the "any but not all gate."

Objective

In this activity, you will demonstrate the operation of logic exclusive OR and XNOR gates.

Materials and Equipment

1–5 V Dc power supply
1–7486 XOR IC
1–7404 inverter IC
1–150 Ω resistor
1–Breadboard
1–LED
2–SPDT switches
1–VOM

Procedures

1. Wire the switches so they can be connected to a breadboard.

2. Insert the 7486 IC into the breadboard.

3. With power off, connect the circuit as shown in **Figure 37-4A.** Refer to **Figure 37-4B** for power and ground connections.

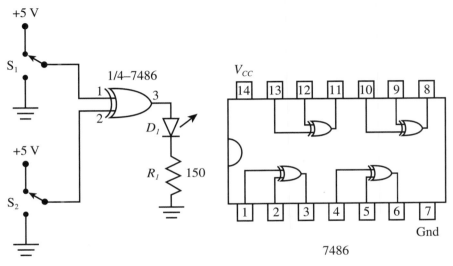

Figure 37-4A

Figure 37-4B

7486

(Continued)

4. Turn the power on. With the input connected to ground for a logic 0 and 5 V for a logic 1, complete the table in **Figure 37-4C.**

Inputs		Output
S_2	S_1	D_1
0	0	
0	1	
1	0	
1	1	

Figure 37-4C

5. Turn the power off.

6. What is the relationship of the input to output logic levels?

_____

_____

7. How is this gate different from the OR gate?

_____

_____

8. Write the Boolean expression for the circuit.

_____

_____

9. Connect the circuit as shown in **Figure 37-4D.**

10. Turn the power on. With the input connected to ground for a logic 0 and 5 V for a logic 1, complete the table in **Figure 37-4E.**

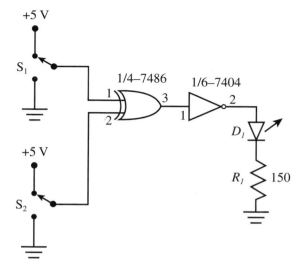

Inputs		Output
S_2	S_1	D_1
0	0	
0	1	
1	0	
1	1	

Figure 37-4D   **Figure 37-4E**

(Continued)

Name _____

11. Turn the power off.

12. Look over the results in the table and decide if the circuit is operating correctly.

13. What is the relationship of the input and the output of the circuit?

_____

_____

14. How is this gate different from the NOR gate?

_____

15. Write the Boolean expression for the circuit.

_____

Summing Up

In your notebook, record any observations, problems, and conclusions for this activity.